A Passenger's Guide to Airline Flying

A Passenger's Guide to Airline Flying

Captain Larry Coy

Pentland Press, Inc.
England•USA•Scotland

The Denver Chart on page 39 was reproduced with permission of Jeppesen Sanderson, Inc. Not to be used for navigation. Jeppesen Sanderson, Inc. © 1998

Figures 1 and 3 were reproduced with permission of *Airliner* magazine, from the article entitled "The 737: Making a Safe Airplane Safer" (July-September 1997).

All other photographs were reproduced with permission of Boeing Commercial Airplane Group.

PUBLISHED BY PENTLAND PRESS, INC.
5122 Bur Oak Circle, Raleigh, North Carolina 27612
United States of America
919-782-0281

ISBN 1-57197-123-8
Library of Congress Catalog Card Number 98-065643

Printed in the United States of America

Contents

Introduction

This book is an attempt to help you, as a passenger, fly more comfortably on a modern jet airplane. By having me share over forty years of flying experience, you will understand more about airplanes, pilots, airports, flight attendants, and the many other factors involved in your flight. You will have a more enjoyable journey, in what is today's safest method of transportation.

My love of airplanes was manifested by soloing a Piper J-3 Cub as a teenager and then flying in the Navy for five years before joining an airline in 1966. Many "hours of sheer boredom interrupted by moments of stark terror," have made me into what one could call an experienced pilot. I am, however, always intrigued by questions from passengers of all ages who don't understand the mechanics of airplanes, government regulations, why the in-flight meal was terrible, why the flight departed late, will an airplane this big actually fly on only one engine, why they can't use their cellular phones, and many more subjects; some for real and some in humor. My goal is to answer the more serious questions about today's airplanes, so that you will be more comfortable on your next flight.

The main concerns of any airline today are making a profit and promoting safety. The airlines of the United States are the safest in the world. Many have attempted to investigate how many hours you can fly today, without an incident. Let's just say more passengers are flying today than ever before, and even with the highest number of departures in history, accidents are at an all-time low. The airlines in this country will continue to become safer every year, taking advantage of the increasing demand for safe, economical, and comfortable air transportation. If they don't continue to improve safety, they will eventually be out of business.

Hopefully, I can explain to you why an airplane flies; how good the pilots are; how well airplanes are built, maintained, and paid for; where the government fits in; how an airline makes money; and then show you the humorous side of the airline business.

I will answer some questions such as: Why does this flight need so many stewardesses just to serve those terrible turkey sandwiches, after which they sit in the back of the airplane to talk about last night's layover? Is that young man with three stripes really old enough to fly an airplane? Can that old guy in the left seat with no hair, really pass a physical exam or should he be in a retirement home or on a golf course? Is that attractive blonde in the captain's seat the senior flight attendant? How are we going to be able to land in Paris if the pilots don't speak French? Will my dog be warm enough down below with the baggage? What happens if the airplane gets hit by lightning? Should the person in the control tower be called an air traffic controller or coordinator? How old is this airplane and what would a new one cost?

As an informed and relaxed passenger, you will realize the airplane you are flying on is a marvel of modern technology, flown by competent professionals, regulated by government officials, and owned by an airline which has a primary goal of providing you with a safe, comfortable flight, that arrives at your destination on time. This book is intentionally small and avoids detailed, scientific discussions about the "magic" of flight. Flying can't be too tough, as the birds have been doing it since the beginning of time. Let me simply give a brief overview without getting involved in all the nuts and bolts of a modern passenger airliner; let us attempt to learn more about how an airplane flies, without becoming too technical.

Chapter twenty-nine is a glossary of technical terms for the reader. It will enable you to obtain an in-depth description of a term, while helping me keep to the basic text and avoid getting tied down in technical aviation talk.

The last page is a method for the reader to ask questions, which I will attempt to answer in future editions. It is also a way to order *A Passengers Guide to Airline Flying* by direct mail.

1

Why Does an Airplane Fly?

Airplanes need wings and engines to fly. There are many other fancy aeronautical items added to make the job easier and more efficient, but basically you need a way of generating lift to get the airplane into the air, which is what the wings do. Therefore, the wings are attached to the fuselage so the whole airplane will fly. We need to push the wings through the atmosphere, so we have engines which produce thrust to move the airplane forward.

There are four forces acting on an airplane in flight. Lift is produced by the wings, relying on a principal of physics first discovered by a gentlemen named Daniel Bernoulli in 1738. Bernoulli's theory states that the difference in the speed at which a liquid passes over a solid, will determine the pressure at that point. Just before the twentieth century began, many inventors were experimenting with different shapes of wings in an attempt to develop lift. Thanks to Mr. Bernoulli, they realized that by developing an airfoil which made the airflow take longer to pass over the top of the wing than the bottom, they could create lift. This happens because the pressure on the bottom of the wing is higher than pressure on the top. The air moving over the top has to travel a longer distance over the curved wing, than the air below the wing which is passing under a flat surface. By making the air travel a longer distance there is less air pressure on top, allowing lift to be created by a difference in pressure.

The first flying machines, which were really gliders, were developed in the early 1900s when Orville and Wilbur Wright brought their flying machine from Ohio to North Carolina to experiment with a new wing, built with wood and fabric. They first glided down Kill Devil Hill near Kitty Hawk. Soon after, they laid an engine on the wing and attached belts to the shafts to move two propellers. This made the machine move through the air faster than simply gliding, and therefore produced more lift. On a December morning in 1903, the first airplane flight produced lift with new wings and moved forward powered by a very basic internal combustion engine, which moved the propellers to produce thrust. An added advantage over gliding with no engine was that they could now also climb with no wind. Men could now fly like birds because of this new understanding in the principle of developing lift by producing more pressure under a wing. The pressure increased as the wing moved faster, being pushed through the air by propellers. The other two forces to be dealt with were drag (which slowed the airplane), and gravity (which pulled the airplane to the earth), but first let's just concentrate on the wings with a short review of what we learned in high school geometry.

A car moves forward and can turn, thus moving in two dimensions. An airplane adds a third dimension by climbing or descending, therefore moving on three different axes. The pilot can increase engine power to make the airplane go forward at a faster speed, move the control stick left or right to make the airplane bank its wings, and change direction of flight by changing the roll axis. If the pilot pulls back on the control column, he changes the pitch axis, and the airplane starts to climb. Keep this axis discussion in mind as we explore the parts of an airplane which actually move to change the flight path. This occurs by moving surfaces on the wings and tail, thus changing one of the three axes.

Small airplanes are controlled with a "stick" between the pilots legs. On a large airplane a "steering wheel" is mounted on the top of the control stick and the name is changed to a "control column." An airplane with two pilots has two control columns, so either can control the airplane's flight path.

By looking at the wing of a modern jet airplane (see Figure 1), you can see the top of the wing is tapered near the front leading edge, back to the trailing edge. We can see the bottom of the wing is flat, and now know that air must travel further to get over the top of the wing than the bottom. Therefore, less pressure exists on the top of wing, so lift is produced and the airplane flies. The main purpose of the wing is to generate lift, but it also provides a method to turn the airplane while in flight. The wings can be completely filled with fuel, which is pumped to the engines via fuel lines, which run through the wings and the fuselage. Additional fuel tanks are built into the lower parts of the fuselage, between the cargo storage areas. Fuselage is aviation talk for the large body of the airplane with the cockpit in front, tail in the back, and all the passenger seats in between. Airlines use the fuselage exterior for advertising, because with a fancy paint job, it becomes a moving billboard.

As you look out the window, during your flight, you see that parts of the front, middle, and back of the wing are moving. This is very normal and necessary for controlling the direction the airplane flies, and for providing more lift at lower speeds by changing the shape of the wing. Just as the Wright brothers watched the first aviators in the fields of Ohio, we have learned about wing shapes by watching birds move their wings to turn and change their shape just before landing. We are still utilizing these principles today by using ailerons to help us turn the airplane to a new heading, and using the flaps to change the shape of the wing to produce more lift when flying at slower speeds for take-off and landing. The ailerons are attached to the back of the wing on the very outermost part.

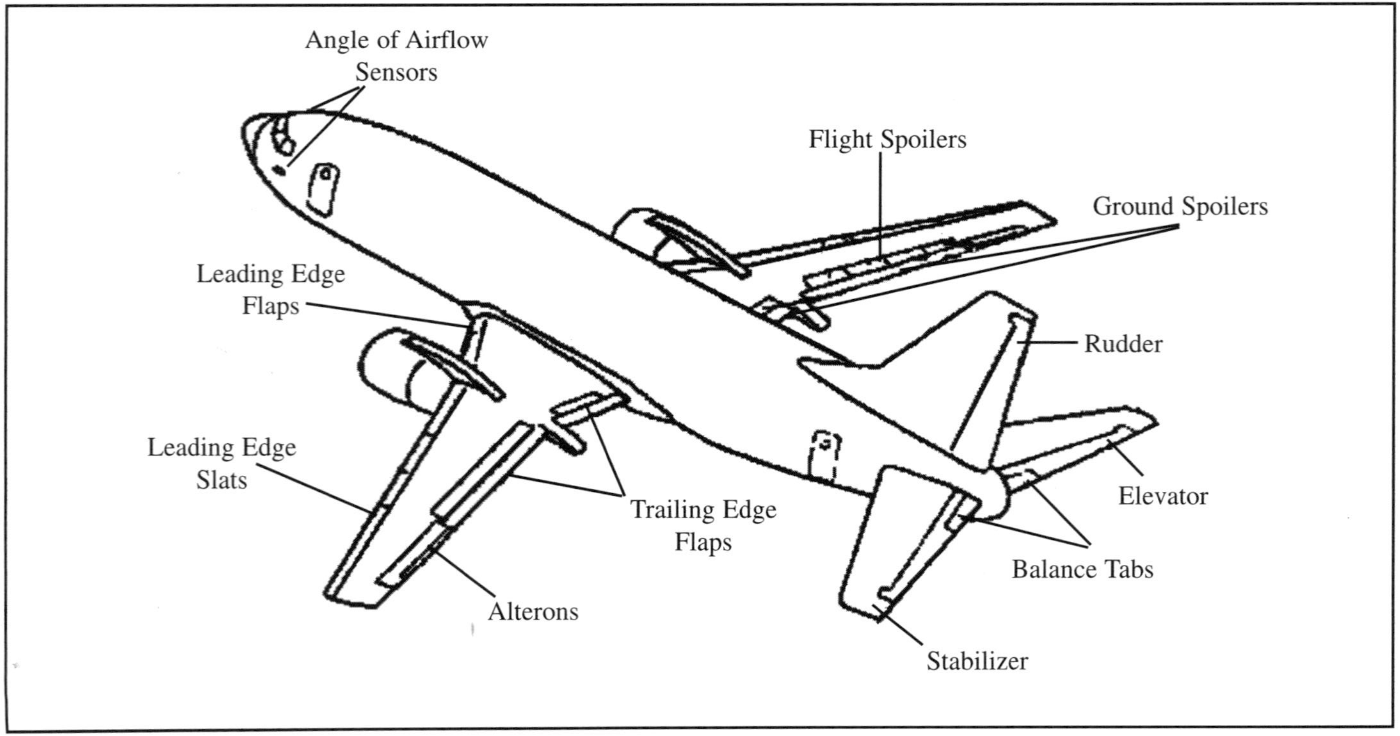

Figure 1
Wings/Control Schematic

They move up and down as the pilot moves the control column left and right. Flaps, on the inner part of the wing's trailing edge, do not move up, but will extend and go downward to change the wing shape for flying at slower speeds. Wings also have leading edge flaps, which change wing shape just like trailing edge flaps do, but the flaps on the front are called leading edge devices or slats.

After landing, you will see panels come up from the middle of the top of the wing. These "speed brakes," or "spoilers," spoil the lift and help the pilot stop the airplane. They also act as speed brakes to slow down the airplane in flight. In flight, speed brakes can help a pilot slow down faster or increase the descent rate. There is also a mechanical connection between the spoilers and ailerons, which moves both simultaneously when the pilot needs to turn faster. So, you may see the spoilers come up at the same time the aileron comes up on the descending wing, when a more rapid turn rate is needed.

Turning the airplane in a different direction is done primarily by moving one aileron up and the other down. Try this experiment to help understand what the ailerons do: as a passenger in an automobile, you can open the window and put both arms outside with one hand right behind the other, as if they were joined. Next, move the back of your rear hand up, and you will feel a force which moves both your hands downward. Then, move the rear hand downward (as the trailing edge of the wing does), and you will notice an upward force on your hands. Next time the airplane turns to the right, you will notice the right aileron is up, making the right wing go down, and the left aileron is down, making the left wing go up, thus turning the airplane by changing the roll axis. The ailerons move in opposite directions; the wing goes up on one side and down on the other, making the airplane turn. Remember, I told you this "pilot stuff" was very basic.

There is another "aileron" on the tail of the airplane, but since this one is mounted vertically, it has been named the rudder, just like on a boat. Just as moving the wing aileron up into the airstream forces that wing down, moving the rudder to the right will force the tail to move to the left, thereby causing the nose of the airplane to move to the right. Moving the rudder changes the yaw axis on an airplane. We can now control what direction an airplane flies by changing roll or yaw. Actually, both roll and yaw must be changed together for a balanced, coordinated turn. You as a passenger should not feel any different forces as the airplane changes direction.

The horizontal part of the tail has a control surface installed on its trailing edge, of which both sides move in the same direction. This is called an elevator, since it goes up and down. To take off, the pilot pulls back on the control column which moves the elevator up, thus forcing the tail down and rotating the nose into the air by changing the airplane's pitch axis. This is getting easier, right?

Our discussion on wings has just about made you into a true aviation expert, except for two other subjects: gravity and drag. In 1589, Galileo Galilei, professor of Mathematics at the University of Pisa, demonstrated gravity to his students by dropping two different-sized stones from the Leaning Tower of Pisa in Italy. In 1670, Sir Isaac Newton was seated in his garden and observed an apple fall to the ground near him, causing him to develop a mathematical theory of gravitational pull. The force of gravity, we all know, is the force which keeps things on the earth or tends to bring them back if they try to escape. The lift we are creating by moving the wings through the air, opposes the force of gravity. Of course, the faster we go, the more lift is created, so it therefore, becomes easier to keep the airplane in the air.

But also, the faster we go, the more drag or air resistance we get. This can also be demonstrated by holding your arm outside the window of an automobile. You can feel the difference when traveling seventy miles per hour on the freeway, as opposed to twenty miles per hour in a school zone. Methods of streamlining moving objects to obtain more speed with less power have continually been improved upon by man, but as we started approaching the speed of sound, drag really became critical. The higher the speed, the more fuel is needed to push an airplane through the air, both for producing lift and for overcoming drag. By streamlining airplanes and reducing exterior parts, we have been able to go faster and do it with less fuel. One example is the retractable landing gear. Remember that lift must overcome gravity and thrust must be greater than drag, for any airplane to fly. For further clarification, there are two types of drag: induced drag, resulting from the production of lift; and parasite drag, such as wheels, rivets, antennas, and windshield wipers.

Now that you are an expert aeronautical engineer, you can begin to understand the art of fine-tuning that continually goes on between finance experts, marketing people, customer convenience sales people, cargo pallet builders, airplane painters, antennae technicians, and lastly the pilots, who know everything about anything dealing with airplanes. Engineers built the huge engines to provide thrust to move the airplane forward, thereby generating lift. The engines also create huge amounts of drag, for which you need to carry more fuel, which of course is needed more than the extra cargo or passengers. Welcome to the world of aviation, which must continually balance fuel requirements, runway length, temperature, weather, cargo, and passengers, against the efficiency of the airplane, while making money to pay stockholders, expensive pilots, and chief executive officers. Airlines must also pay for the airplanes, fuel, and that tasty cold turkey sandwich on a dry roll. All this comes from that very low-priced airline ticket the passenger has to buy. Next, we shall learn about the engines needed to push the airplane through the atmosphere to create lift.

2

What Makes it Go?

A passenger once came into the cockpit prior to departing the gate, and asked how we got the power from the wheels to the wings for take-off, while going down the runway. I pointed to the two rudder pedals in front of my seat and explained that they were clutches, which transferred the power when pressed at rotation speed. Following a few seconds of thought, while staring at the pedals, a smile finally broke out as he realized there must be another explanation. The only power that goes to the wheels is hydraulic power, which is used for stopping and turning the airplane on the ground. The thrust necessary to propel the airplane is developed by the engines, which push the airplane forward by using a propeller mounted to the engine crankshaft. In a jet engine, the combustion inside turns a turbine blade in the back, which is connected to a compressor in the front of the engine, similar to a propeller (see Figure 2).

From the earliest flight at the turn-of-the-century, the source of thrust was a propeller mounted on the engine. By changing the rotation speed and the twist of the propeller blades, a pilot could determine how much thrust he needed at any given moment. A need for more speed in military airplanes

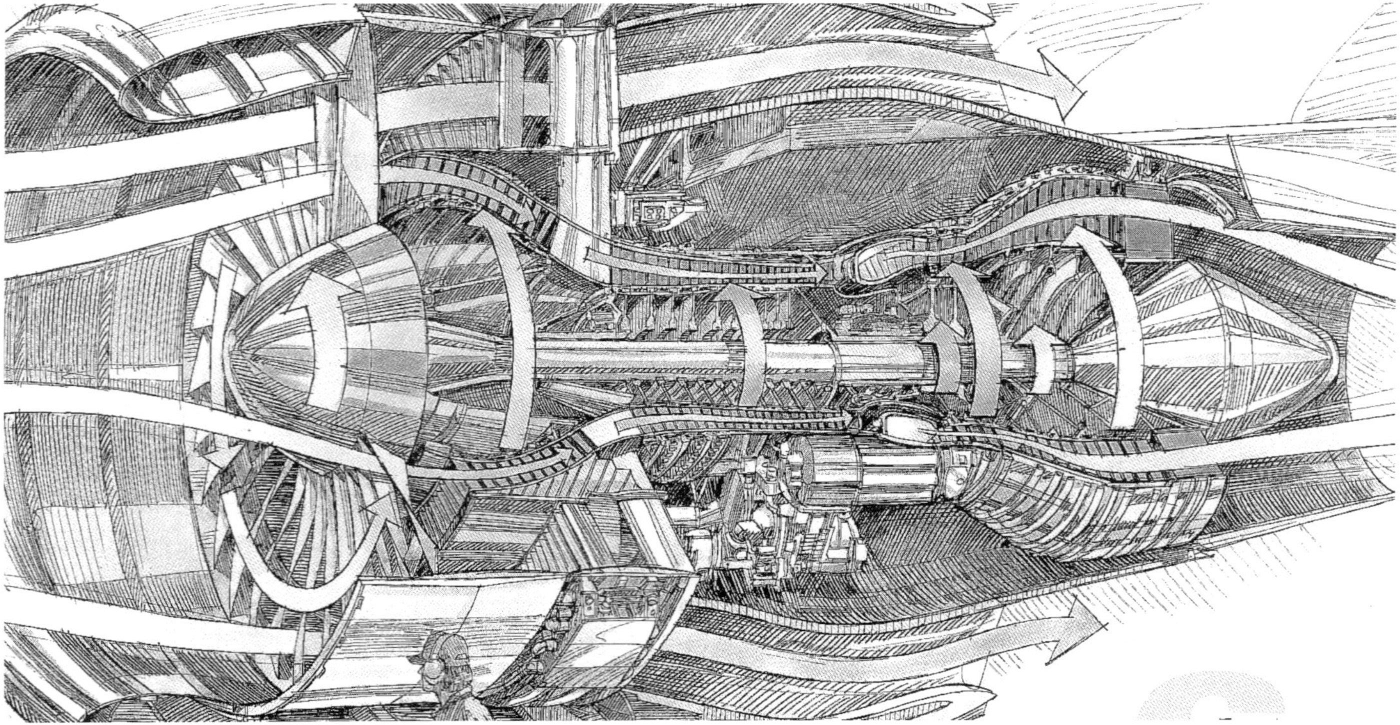

Figure 2
Engines

led to the development of jet engines during World War II. Engines using propellers were internal combustion engines, similar to the engine in your automobile. Jet engines utilize turbines, which are rotating fans mounted on a shaft. Turbines were used in steam engines prior to jet engines. Engineers simply found ways to pull air from the front of the engine, combine it with fuel, and then ignite it in what we call burner cans, which push the airplane forward and exhaust the unused fuel/air mixture out the rear of the engine. They realized that by injecting more fuel into the burner cans, they could turn bigger fans faster and make more air pass through the engine, thereby making it produce more power. To get just a little more technical, the fans you see inside the front of engines are called compressors because they are used to compress the air coming into the front of the engine. There is another set of fans in the rear of the engine called the turbines. The jet engine produces thrust by combining fuel and air in the burner cans that exhaust over the turbine blades. These are connected via a drive shaft to the compressor, which pulls more air in the front. The jet engine (with fewer moving parts), is much more basic than an internal combustion engine. Because it operates at much higher temperatures, new metals had to be developed before more thrust could be produced from newer and bigger jet engines.

While the engine provides the thrust to propel the airplane forward, it also has several accessories attached, such as generators to provide electricity, hydraulic pumps to provide pressure for hydraulic systems, and a system for extracting compressed air (before it is mixed with fuel), which is used to provide pressurization and temperature control in the airplane during flight.

Airplanes make a lot of noise right after landing, when the engines sound like they are reversed. Actually, the engines produce thrust just as they do in flight, and are not reversed. The thrust is simply blocked from exiting the rear of the engine by doors which direct it forward and out the sides, thereby helping the pilot produce reverse thrust to assist the brakes in stopping the airplane. The term "reverse thrust" originated with propeller powered airplanes, which actually reversed the pitch of the propeller blade to direct the thrust forward. Pilots use reverse thrust to slow the airplane's forward movement, but they can also move the airplane backwards on the ground by using it as well. This procedure is called "power back." Hopefully, your flight crew will advise you when they are going to use this method to move away from the terminal, because suddenly there is a lot of noise as the airplane moves in reverse.

While the power output of your car engine is measured in horsepower, the output of the jet engine is measured in pounds of thrust. As an example, the thrust of an engine on a Boeing 727 or McDonnell/Douglas MD-80 is about 15,000 pounds, depending on the engine type. A DC-10 or 747 has engines which develop about 50,000 pounds of thrust each. The new Boeing 777 has two engines, which each produce 90,000 pounds of thrust.

When a military jet fighter takes off, fire comes out of the tail. This doesn't happen with a commercial airliner, with the exception of the Concorde SST. The fire coming out the exhaust is caused by the afterburner. "After burning" is just like it sounds. Engineers have found a way of producing more power by injecting more fuel just aft of the turbine section, producing more thrust by what the British call "re-heating" the exhausting fuel/air mixture. This provides the extra power needed to increase speed for fighter planes, and allows heavier take-off weights. Engines on commercial airliners produce sufficient power for take-off, even if one of the engines fails. Hopefully, we won't need more speed to elude enemy fighters approaching out of the six o'clock position. We also don't need speeds faster than sound, as we would in the supersonic Concorde.

3

Fuel Provides the Thrust

While flying as captain of a Boeing 727 on a flight to Chicago's O'Hare Airport, I made the decision that, due to weather conditions, we had to divert to an alternate airport. In conversations with the company and the dispatchers, it was decided that the new destination would be Milwaukee. There we would refuel and then return to Chicago. A few minutes later we were landing at General Billy Mitchell Airport in Milwaukee and taxiing to the terminal. My airline did not fly into Milwaukee at that time, so another airline was paid to service our airplanes when we diverted there. Unfortunately, we soon learned these folks were carrying picket signs in front of the terminal and not available to put fuel in my plane. I had my co-pilot ask where we might get fuel, and the ground controllers directed us to a place that served smaller, general aviation airplanes but could handle a 727. Upon arrival, we shut down the engines, made sure the tires were chocked—blocks in front and back to keep the airplane from rolling—and informed the passengers that since there were no jetways, I was about to exit the back of the airplane in an attempt to get fuel for our return to Chicago. I exited the airplane below the tail, via the stairs built into the 727, and was met below by three fuel suppliers from different companies, all suggesting they would love to fill up a 727. I jokingly asked who would take a credit card and who had green stamps. We departed for Chicago with my American Express bill showing an expense for 6,000 gallons of jet fuel and 30 books of green stamps to be divided among the crew. Fuel is one of the major expenses of running an airline today, and estimates are that a change in price of one penny per gallon, will change an airline profit for the year by a million dollars. Airplane fuel is slightly more refined than kerosene, which basically is what is used in military fighters. The per gallon cost of jet fuel is about two-thirds of what you pay for gas in your car, but a lot more is burned flying a heavy metal airplane close to the speed of sound, than you use motoring down the freeway at sixty-five miles per hour.

Previously, we learned that fuel is pumped into the "burner cans" in the engine, after being routed from the fuel tanks in the wings and the bottom of the fuselage. There, it is mixed with a spark from the engine igniters, like spark plugs, which are powered by the electrical system. This combustion produces thrust. Technically speaking, there are three problems which must be dealt with regarding this fuel flow. The fuel must be at the correct temperature and pressure when injected, and it must be very clean. Because it has cooled down inside the wing, the fuel is heated to avoid being like jelly, and too rigid to pump through the fuel lines and into the engines. Then it is pumped at increasing pressure into the engine area, and filtered to make sure it arrives very pure and at the correct pressure. Almost all of the fuel is consumed in combustion. Very little of the exhaust you see coming out the rear of the engine is wasted fuel.

In an emergency situation, a pilot can choose to jettison fuel to make the airplane lighter for better flight characteristics and also, to get the airplane weight down to the maximum weight for landing. On most airplanes, there are pipes extending out the aft tip of the wing where the fuel can be jettisoned, or dumped. If the fuel system is operating properly, fuel can be jettisoned at about one thousand gallons per minute, depending on airplane type. Even though this fuel is being dumped into the atmosphere very near the engine exhaust, there is no danger involved. The fuel is being dumped in a vapor state that will not catch on fire and will not reach the ground. Fuel dumping is very rare, and when a crew advises Air Traffic Control that dumping is necessary, they are assigned an area and altitude far from other airplanes.

The bigger the airplane, the more fuel is used. Developing the thrust to lift the wing, thereby getting the airplane into the air, requires power directly proportional to the airplane's weight. As an example: a Piper Cub will use about five gallons of fuel per hour, a Boeing 727 will use about 1,500 gallons per hour, and a Boeing 747 will use about 3,000 gallons per hour. Since airplane weight is one of the most critical parameters of flying, we now measure the amount of fuel in pounds instead of gallons. Jet fuel weighs about 6.7 pounds per gallon, therefore, the jumbo 747 is getting about 20,000 pounds lighter every hour of flight. We don't care how many gallons are on board, except when we receive the fuel bill and when we tell the passengers how much gas we are using compared to their automobiles. Only the number of pounds being burned and what remains is important.

The amount of fuel put on an airplane is planned precisely to get the airplane where it needs to go, and get all the passengers and cargo on board within the limitations of take-off and landing weights. Then the captain adds a little bit more to make him feel better or to go faster on his homeward journey, all of which is agreed to by the flight dispatcher and load planner, but without permission from the airline finance department, who has to pay the bill.

Just as the cargo is placed in certain parts of the lower fuselage to keep the airplane balanced, the fuel is loaded on board in a precise manner to meet center of gravity (CG) specifications for that given flight. To maintain the correct CG and assist the airplane, the fuel is burned first from the fuselage tanks and then out of the inner wing tanks. This keeps the wings "loaded," with the heavier part of the wing on the outside for wing structural design purposes.

The wing is a very complex fuel tank, with flaps and ailerons on the back, spoilers on top, slats on the front, static discharge wicks on the trailing edge, hydraulic lines going to flight controls, lights for multiple uses in different locations, maintenance access panels everywhere, fuel dump pipes on the back, emergency escape paths on top, and engine pylons supporting the weight of the engines. But the most important function of the wing is to generate lift while containing thousands of gallons of fuel.

When in Vietnam, I purchased a Honda 50cc cycle for transportation. Rather than buying gas, I found it much easier to simply park under the wing of an airplane and drain a few gallons into my small tank, rationalizing that Uncle Sam wouldn't miss a few drops. Prior to my final departure from that beautiful country I sold my cycle, realizing that it would really go fast for a 50cc, and hoping that the motor had a few good miles left after burning AvGas (Aviation Gasoline), instead of the regular eighty octane from the gas station. I'm afraid that the Vietnamese gentlemen I sold the cycle to may never trust a Yankee again.

4

Oxygen Masks in the Ceiling

If you feel a sudden pressure change in your ears, possibly followed by a loud noise, you will probably see a small plastic object, dangling on a clear plastic line, fall out of the airplane's ceiling right in front of your nose. Don't ask questions or look around— put the oxygen mask on and then assist the person flying with you, because the airplane pressurization system has a problem. This situation is extremely unlikely, but it is the reason that the flight attendants must demonstrate this system to you prior to every flight. Put the mask on over your mouth and nose while placing the rubber holder behind your head to keep it on, then continue wearing it until advised by a flight crew member that you can remove it. If the situation was caused by a real problem, the cabin will get colder, visibility will become reduced, and the noise level will increase due to increased air flow and passengers exercising their vocal cords. Shortly after this, you will notice the nose of the airplane going down as you start a descent to a lower altitude, where there is sufficient oxygen to breath without pressurization.

To help you understand the pressurization system, simply realize that the structure of the fuselage is just like a balloon that is constantly being inflated. Air is taken from outside the airplane and mixed with air from the compressor section of the engines. The air is then passed through ducting into the cabin before being vented to the airplane's lower sections, where the baggage, cargo, and animals ride. Outflow valves located on the bottom of the fuselage are controlled from the cockpit to regulate air leaving the airplane. This maintains a controlled amount of pressure within the cabin. This process results in circulated air being completely changed in the passenger areas about every seven minutes. The pressurization and temperature control systems are usually controlled automatically, but have manual back-up systems which the pilot can elect to use at any time.

The temperature is controlled by mixing the cold outside air with heated compressed air taken from the engine in mechanical devices called "packs." The air is then delivered to one of many overhead outlet areas throughout the cabin, at a temperature and pressure selected by the pilot.

Should there be a need for oxygen from your little yellow mask, in most airplanes it will come from oxygen cylinders in the airplane's cargo area. In some newer planes, the oxygen would be generated chemically in the back of the seat in front of you, and this would be evident when the oxygen mask comes from the seat and not the ceiling. Flight attendants and pilots are also provided with separate portable oxygen systems, which they can use by carrying smaller containers with masks attached. This enables them to make their way through the cabin and help you during an emergency.

Airplanes are pressurized because the human body needs oxygen. While there are sufficient amounts at the earth's surface and lower altitudes, as the airplane climbs there are diminishing amounts of oxygen available. Airplanes today fly at altitudes of nearly 40,000 feet and air temperature decreases approximately three degrees Fahrenheit for every 1,000 feet. Therefore, breathable air at the correct temperature, supplied by a dependable pressurization system, is necessary for crew and passenger comfort and safety.

Sometimes you board an airplane in the summer, the cabin temperature is terribly hot, and the pilot advises you over the passenger address (PA) system that the APU (Auxiliary Power System) is not working, or the ground personnel have not plugged in the GPU (Ground Power System). This is aviation talk which means that they can't get the auxiliary power unit started or the ground power unit is being used by the airplane at the next gate. Either of these units can provide air to the airplane and, in fact, are needed to start the engines.

Auxiliary power units (APUs) are small jet engines which are built into the tail of the airplane to provide auxiliary electric and air power for ground use, such as starting engines, providing hydraulic power, getting air into the cabin at the correct temperature, and most importantly, getting power to the galleys so the flight attendant can make coffee in the morning. They can also be started in the air should an engine fail or not supply sufficient electric or pneumatic pressures for airplane systems, but they will not provide thrust for the airplane.

Most airplanes also have a device called an "air driven generator," which can be dropped into the airstream while in flight, if the need arises. It has a small propeller which will turn in the air stream to provide some power, such as electricity, hydraulic pressure and possibly some other emergency needs.

Pressurization problems are usually caused by temporary mechanical problems or pilots putting incorrect information in the controllers. Neither will result in masks dropping from the ceiling, simply an unhappy flight attendant hurrying forward to the cockpit to voice her opinion concerning the ability of the pilots to understand how to operate the system. A major problem is caused when there is a structural failure of some part of the airplane, such as a cabin door or window departing from the fuselage. The pressurization system is designed to provide a sufficient amount of air into the cabin, but a sudden increased demand for air, because more air is being exhausted via a new hole in the airplane, makes the demand more than the system can provide. A change in air flow will be noticed, and the oxygen masks will drop in front of you. Anything that is loose in the vicinity of the new hole in the fuselage will suddenly leave the airplane prior to the actual arrival at your destination. This is a very good reason to have your seat belt on at all times, because it could be very embarrassing, not to mention painful, to suddenly leave an important business meeting in the air, in favor of making an uncontrolled dive out of the airplane to a very hard surface below, leaving your luggage behind.

These systems, as with the engines and wings, are technically very sophisticated and very dependable. But like anything mechanical, they are subject to fail as discussed in Murphy's Law, which describes when and how things quit operating correctly. Please pay attention to the flight attendant's demonstration while taxiing out for take-off. It helps their ego if they know you are looking at them, but also so you will realize the importance of knowing what to do and what actually causes the little yellow mask to appear, as opposed to the passenger next to you who didn't have a seat belt on and who suddenly excuses himself.

It is interesting to see an accidental deployment of oxygen masks and note how many passengers simply disregard the mask dangling in front of their nose. Some who do put them on, keep moving their heads from side to side, just like the flight attendant did in the demonstration.

5

Taking Off

The time from when you hear the engines increasing the noise level to full power, until the pilot pulls back on the control column when enough speed has been reached for take-off, is usually less than one minute. Planning for this take-off started many years before, when the airplane was initially designed. During the last few hours prior to actual departure from the runway, many different people have been inputting information that will lead to a safe, comfortable departure to your destination. The airplane manufacturer, in concert with the airline company, under the supervision of the Federal Aviation Administration, has designed an airplane that can lose the most critical engine or encounter any myriad of other problems, and still get safely into the air. Pilots and crew are trained to deal with any situation and return the airplane safely to the airport.

On the ground, maintenance personnel check the airplane completely, as well as the pilots who will be flying with you. Weather people advise the flight crew about conditions such as temperature, winds, and precipitation possibilities, all of which have to be considered to determine the aircraft weight, which runway to use, and an alternate departure airport should you be unable to return to your original airport in an emergency. Load planners have determined how much cargo can go along with your baggage, if all the seats can be filled with passengers and still be within take-off weight limitations, and where to put Fido and Boots down below. All of this is achieved, as well as an agreement between the company dispatcher and the pilots for the correct fuel load, to get you to your planned destination or an alternate airport, plus additional fuel for holding when ATC (Air Traffic Control) decides they need more space between arrivals. The Postal Service would also like to load a few thousand pounds of mail. Most of the mail delivered in this country is flown below you in the cargo compartments.

After all these folks agree, the airplane can safely prepare to depart the gate area. We then start dealing with a government agency, the Federal Aviation Administration, which controls departures from all airports. They make sure there is enough room in the sky for all the airplanes to fit into slots on airways, which are highways in the sky. We then taxi out to the runway being used for departures. All of this takes very little time if you're departing from a small airport for a connection at a larger airport, but after you make your connection at the airport with hundreds of departures every day, it seems just one little problem can delay your take-off for minutes, if not hours. Your captain should keep you informed concerning all aspects of your flight. If you have a question, ask an agent or flight attendant to get an answer for you.

In the cockpit, the pilots agree that the aircraft take-off weight, weather conditions, and runway are clear to proceed. Checklist items have been challenged and answered to be sure all instruments, lights and switches are functioning correctly. Flight attendants confirm the number of passengers on board. The ground crew is advised that the airport ground controllers have given their permission to leave the gate and start the engines. While the push-back crew is moving the airplane backwards, you will notice the air conditioning has stopped. The pilots need these same air sources to turn the engines just before they introduce fuel and ignition. The co-pilot then calls the tower for taxi permission, and the captain increases power to taxi to the runway while steering the airplane with the nose gear. While you are taxiing out, the co-pilot is using a check list to make sure all items, such as flaps, radios, take-off speeds, de-icing equipment, and many other items, are set correctly in concurrence with the captain. As you get close to the runway, the radios are changed from ground controllers to tower controllers, who will give permission for take-off in coordination with landing airplanes. During the taxi, the flight attendants have been doing their safety briefing, to which you should have been paying attention.

For the actual take-off, the pilots increase the engine thrust to full power while keeping the airplane aligned with the runway centerline. Engine instruments are monitored for correct power settings and malfunctions. When VR (Rotation Velocity) speed is reached, the pilot pulls back on the control column, rotating the nose up, and the airplane leaves the ground. Next, you hear the landing gear being raised into the fuselage, and shortly after, the flaps on the wings will be retracted for flight at higher speeds. Then as you pass a thousand feet, you will hear a slight power reduction as the engines are reduced from take-off power to climb power. Shortly after take-off, you will notice the nose is lowered and the airplane speed seems to increase. However, our speed is limited to two hundred and fifty knots until we climb above 10,000 feet. After this, we will climb at higher speeds until we achieve our Mach speed, which is eighty-five percent of the speed of sound. We will discuss speed in greater detail later on.

A rejected take-off is when the airplane has started the take-off roll and for some reason does not take off. This is the leading cause of aircraft accidents today! A take-off can be rejected for any reason that the pilot feels it is unsafe to continue, such as a mechanical problem, changing weather conditions, a control tower ordered abort, or a flight attendant calls to say the lady in seat 15B wants to get off. Another speed involved here is V1 speed, which technically is the speed at which the crew has decided, based on runway length, that the airplane cannot safely be stopped on the runway. Therefore, after reaching V1 speed we are going to fly, because it is safer to fly than to stop on the remaining runway. Actually, V1 speed is determined from manuals and computers based on airplane weight and a few other factors, but mostly on runway length. V1 speed is usually about ten knots below VR speed, which is then about ten knots below V2 speed, which is the climb speed that is flown for the best airplane performance with an engine not operating. That's all the technical on the ground speed stuff that you have to know for the test.

The reason the rejected take-off is causing so many problems is that things are happening in a hurry. Every situation is different, and the pilots have to make very rapid decisions concerning the airplane stopping on the runway or continuing the flight. If they abort the take-off, then they must decide whether to get off the runway or remain there where other flights cannot take-off or land. They then must decide if the situation is bad enough to evacuate the airplane. All of this is complicated by the tires and brakes being very hot and the possibility of either an engine or another system being on fire. Smoke is coming from someplace, everyone wants to know what's happening, fire trucks with flashing red lights are on the way, and the crew wants to get home for Monday night football.

Should you ever be involved in a situation where the airplane stops before taking off, do not make any decisions by yourself. Flight attendants are well-trained to let you know what to do, and the last thing we want is a solo passenger starting a panic by opening a cabin door and deciding he wants out. The captain will let you know what's happening as soon as possible, and if you are supposed to evacuate, you will be told in very loud voices where to go. Pay attention and move as quickly as possible, but forget about your lap-top computer and your carry on baggage.

As the take-off weights on today's newer airplanes approach one million pounds and new engines are developing over 100,000 pounds of thrust, we are seeing bigger airplanes holding more fuel, and taking more passengers and cargo to newer destinations, with a vast increase in safety and passenger comfort. Sit back, enjoy your flight, and realize that the state of technology and professionalism of the flight crews involved are making your trip a real pleasure, and a bargain when economically compared with any other form of transportation. And now you know more about what is actually happening up there behind the closed cockpit door.

6

How Fast are We Going?

You are going 650 miles per hour, 280 knots, and Mach .84, all at the same time. Earlier we discussed the speeds used on the ground for take-off, which are very similar to speeds used in the automobile, except the speed in your car is determined by how fast the wheels are turning, and in an airplane, the speed is measured by air entering an open-ended tube which can be seen just below the cockpit. The "pitot tube" determines how much air pressure is exerted on the open end of the tube and converts this into "indicated airspeed." This ram air pressure shows on the pilot's airspeed indicator how fast the airplane is moving through the atmosphere. While your car is affected by wind, it is never measured and is not as significant as the winds are to an airplane. You can be moving through the air at one speed but your speed over the ground can be significantly different, depending on the winds. If you are flying into a fifty knot headwind, your ground speed will be fifty knots less than your indicated airspeed. This is why west-bound flights take longer than east-bound flights; i.e., the pilot flies the airplane at the same speed through the air, but most winds at altitude are out of the west, so your ground speed can be two hundred knots different. Flights across the United States, from coast to coast, vary in flight time up to two hours, depending on whether you are going to New York or San Francisco.

You may think the captain is trying to impress you by saying the speed is Mach .84, which is 84% of the speed of sound, but actually this is the speed terminology used by pilots and controllers to achieve a common standard in comparing one airplane's flight path to another at the higher cruising altitudes. The speed of sound varies with temperature and air density, so while sound travels approximately 650 miles per hour on the surface, at 40,000 feet, the speed of sound is down around 450 miles per hour. Making an airplane go supersonic (faster than sound), involves higher indicated speeds at lower altitudes. A lot of this is pilot talk and is not very important to a passenger, except when they want to impress the flight attendant with aviation lingo.

I was once traveling as a passenger in the first class section behind the cockpit of a Boeing 747, in the hump above the regular first class section. I overheard two aviation experts (after they had a few drinks), discussing their aeronautical knowledge with each other. One gentlemen was explaining that he could always tell if the airplane was going west or east because of the sound of the airflow over this part of the airplane. He claimed that when the winds were out of the west, the wind noise was higher, to which the other gentlemen agreed the noise level did seem higher today when going toward the west coast. But you now know the airplane is going to move through the air at the same speeds regardless of direction, unless this is the final leg of a four day pairing for a flight crew that is going home. Tail winds are just an additional positive factor to increase speed over the ground, but wind does not change the airplane's "indicated air speed" or how fast the airplane goes through the air.

You will also notice that as you get closer to the ground upon arriving at your destination, the airplane starts slowing down. Remember that our speed is limited to 250 knots below ten thousand feet, on approaches as well as departures. Air traffic control usually assigns even slower speeds to make our arrival compatible with other flights. We also need to slow down to avoid exceeding limit speeds on the landing gear and wing flaps, as they are built for only so much air resistance. We will be landing at the slowest speeds possible to save wearing all the rubber off those big tires, and meeting the limitations for getting the airplane stopped within our assigned runway's length.

To get you to your destination on time, cruising speeds are near the speed of sound, but slower speeds, for landing and take-off, must consider VS, or "stall speed." This speed has nothing to do with the engines stalling, as in a car. The stall speed of an airplane is that speed at which not enough air is passing over the wing to provide enough lift for the airplane to fly. Technically, the wing stalls when airflow over the wing is decreased to the point that sufficient lift is not produced to support flight. This causes the airplane to become very difficult to control and pilots must correct the situation by adding power and pushing the nose over to increase speed. Pilots practice stalls in simulators during annual proficiency checks. The pilots will be touching down on the runway at a speed which is 30% above the airplane's stall speed.

How fast are we going? Do we mean through the air, over the ground or in relation to the speed of sound?

7

Where are We?

Following the development of bigger and more advanced airplanes during the First World War, it became obvious that these airplanes could be used for delivering mail instead of bombs, at a faster speed than existing surface transportation. The only problem was that trains simply followed the tracks to the next town, but pilots didn't know where the city was at night or when the winter snows started. The government assisted by installing transmitting beacons over the country, so airplanes fitted with receivers could monitor these transmitters and home in on their locations, which were generally at an airport. Rotating beacons, or those green and white lights you see revolving on top of the control towers, were also installed to help pilots spot airports from longer distances. These systems were adequate for airplanes carrying mail, but a few years later when the first passenger-carrying airplanes were introduced, the safety factor had to be increased. A single-pilot airplane with a few sacks of mail could still land in a farmer's field if he got lost, but this wasn't safe or practical with a dozen passengers on board.

Several new lighting systems, such as runway lights, were installed at bigger airports to help pilots see the runway after finding the airports at night. However, better electronic navigational systems were needed and invented. More powerful transmitters were installed at more locations all over the country. Then, when another war began in Europe and Asia in which the United States soon became involved, we had to learn how to navigate airplanes over the oceans, because beacons could not be installed over water.

While land-based beacons were being perfected for short-range use, new systems were also being developed for long-range flights, such as LORAN, or long-range navigation, for which towers were installed on the shorelines. There were transmitters located on coastal land sites that transmitted continuous signals over the oceans. Directional beacons also extended out into the ocean. This assisted pilots as they got closer to land.

Enough history for now, but simply understand that just as airplanes were being modernized from the initial fabric-covered flying machines of World War I, methods of navigating were also being developed to help pilots fly these new airplanes all over the world.

As you are driving across the country, in many places you will see white towers about thirty feet high and twenty feet across which get narrower toward the top, which is flat. These are VOR (Variable Omni-Range), transmitters, which give pilots both directional and distance information. Where the original beacons only provided directional information via a needle pointing at a compass card in the cockpit, the VOR towers also provide distance information telling the pilot how far he is away from the tower, up to two hundred miles. Each of these towers has a separate frequency, so the pilot knows exactly which tower his radio is listening to. This has been the primary source of airplane navigation until very recently.

Several years ago the Defense Department started placing satellites in orbit to transmit navigational information to our military forces. These systems are extremely accurate, and their usage completely eliminates the need for land-based navigational systems, except as back-up units. Very recently, commercial airlines have been installing receivers of satellite information on airplanes for more accurate navigation and also more economical direct route flying. Direct-route flying avoids going to each navigational facility and allows the crew to navigate directly to the destination, thereby decreasing miles flown. This does not affect your frequent flier miles, but it does save fuel.

Basically, there are twenty-four satellites in orbit around the earth, of which airplane navigational systems use at least three to triangulate the electronic information into a very accurate position. The satellite systems are called GPS (Global Position Satellites), which can also provide additional information, such as an airplane's altitude and distance from other airplanes. These systems are being developed today for use in the near future. Very soon all airplanes will be relying completely on satellite-transmitted information to fly all over the world, with much more accurate navigational information.

Navigational systems, which are used to help a pilot land, are somewhat different, but also are slowly being changed to satellite information to find the correct landing location. Since the 1950s, most pilots have used an "instrument landing system," or ILS, to find the runway. The ILS gives directional information to the end of the runway, but also provides altitude information to the pilot by displaying glide slope information, or showing how far the airplane is above the runway as the airplane continues its descent for landing. Most ILS systems also provide "distance from the runway" information. Using an ILS, the pilot can descend to as low as one hundred feet without seeing the runway and still land safely in poor weather conditions. Today's newer airplanes also have an automatic landing system which is referred to as Category Three landing, in which the airplane, which is equipped with enough electronic receivers and has all the other systems working correctly, can actually land itself while the pilot simply monitors the systems and turns off the runway after the airplane has landed.

Additionally, all airplanes are continually monitored in flight by air traffic controllers from the Federal Aviation Administration. All airports have controllers in the tower who view airplanes on the ground until take-off. Then they transfer control to departure controllers who use radar screens to track the airplane's ground movement as well as altitude information. Eventually, control of the airplane's movement is turned over to air traffic control centers, about ten of which are located around the country. While not a navigational system, by monitoring radar scopes, controllers advise pilots about their positions, other airplane locations, approaching weather systems, and usually use stronger language to inform a pilot he is not going in the correct direction or at the proper altitude.

Longer-range navigation for airplanes flying over the oceans has depended on "inertial navigation systems," (INS), of which three are usually installed in the airplane. INS can determine the airplane's position by simply measuring wind information along with heading and ground speed. Inertial systems are being replaced by the much more accurate GPS systems.

Foreign airlines are left to modernize their own navigation systems, mainly because satellites were put into orbit by the U.S. Defense Department, and the technology flow has been easier to provide to U.S. airlines.

The growth projections for air travel are tremendous, and the newer navigational systems are very necessary to tell pilots their exact locations and also to provide more accurate separation between airplanes. More fuel savings are possible because of more direct-route flying.

You should feel very comfortable flying as a passenger today because the navigational systems can tell the pilot exact location information, which is also continually monitored by air traffic controllers, who also give permission for more direct-route flying. Some airplanes on long-distance flights display a map showing your location, speed, and altitude on a movie screen in the cabin or the individual seat TV monitors. This data is taken directly from the equipment the pilots use for navigational information. If you still want to know more, simply ask a flight attendant to ask the pilot if he would mind announcing what river you are flying over, as there are probably thirty other passengers that would like to know as well. The pilots will dig out the Rand McNally Road Atlas and exclaim to each other that they never knew the river's name, then make an announcement which makes you feel like they are intimately familiar with every turn in the river.

Later in the book, there is a chapter which goes into more detail on the subject of navigation. To know your exact location on the Earth, you must know your distance from the equator, or your latitude; the longitude position tells you how far east or west you are while going around the Earth. A bit of historical perspective is included.

8

Jeppesen

After World War II, pilots had little to rely on for navigation except road maps. New navigational facilities were being built around the country and could be added to their charts. Airlines depended on pilots to devise their own methods for getting to a destination and deciding what to do if they couldn't find the right airport. An enterprising pilot by the name of Jeppesen decided he could make a lot more money in the publishing business than as an airline pilot, and he could be home every night. Realizing that things were changing with a new government agency called the Federal Aviation Administration, and that air traffic would be growing tremendously, Jeppesen saw a need to standardize air navigation with maps for air travel that didn't rely on looking for train water towers or truck stops.

All airlines in the United States today issue Jeppesen charts and approach plates to their pilots. They are updated twice monthly to show new information, such as frequencies, runway length, lighting, and anything else vital to a pilot flying into an airport. As navigation technology has changed the need to continually update maps, the Jeppesen Company has met the challenge and provided pilots with information about airports and cross-country navigation. All this is done with the Federal Aviation Administration overlooking changes. Pity the poor pilot who is on a flight when a guest FAA, or Federal Aviation Administration, inspector shows up and discovers the charts in the pilot's flight bag are not up-to-date.

As a kind of review, I have included a chart (see Figure 3), for the new airport at Denver, and I feel that going over some of the information contained on it would be helpful in understanding what we have already covered in this book. Keep in mind that most major airports in the country have about twenty pages of information, so the standard Jeppesen domestic airport manual is about three inches thick. Denver Airport is a good example, since as sports fans know, the city is a mile high. This lets us explore the differences between sea level altitude and actual heights above the ground. It is also the home of the Jeppesen Company.

Starting at the top of our Jeppesen chart, we can see the number of this chart and the last time it was updated, and that it is for Denver International Airport in Denver, Colorado. Just below, we can see this particular approach plate is for an ILS approach to runway thirty-four, which heads nearly north. Runway 18/36 would be north/south exactly. Below this we see Category II and III which means these are the approaches used for bad weather when the snows come to Colorado. We know that Category III means the airplane lands itself with the pilots monitoring the systems. Next is KDEN which is the international identifier for this airport. All airports in the U.S. start with a "K." Then it says the airport elevation is 5,431 feet with the runway touchdown zone (the area where the airplane will land), being 5,351 feet or eighty feet lower due to the changing terrain and the fact that the airport elevation is always measured from the center of the airport. Now look at the three blocks on the left of that line which show that the correct frequency for an ILS approach is 111.1 on a course of 350 degrees and intercepting the glide slope at 6,963 feet or 1,612 feet above the surface.

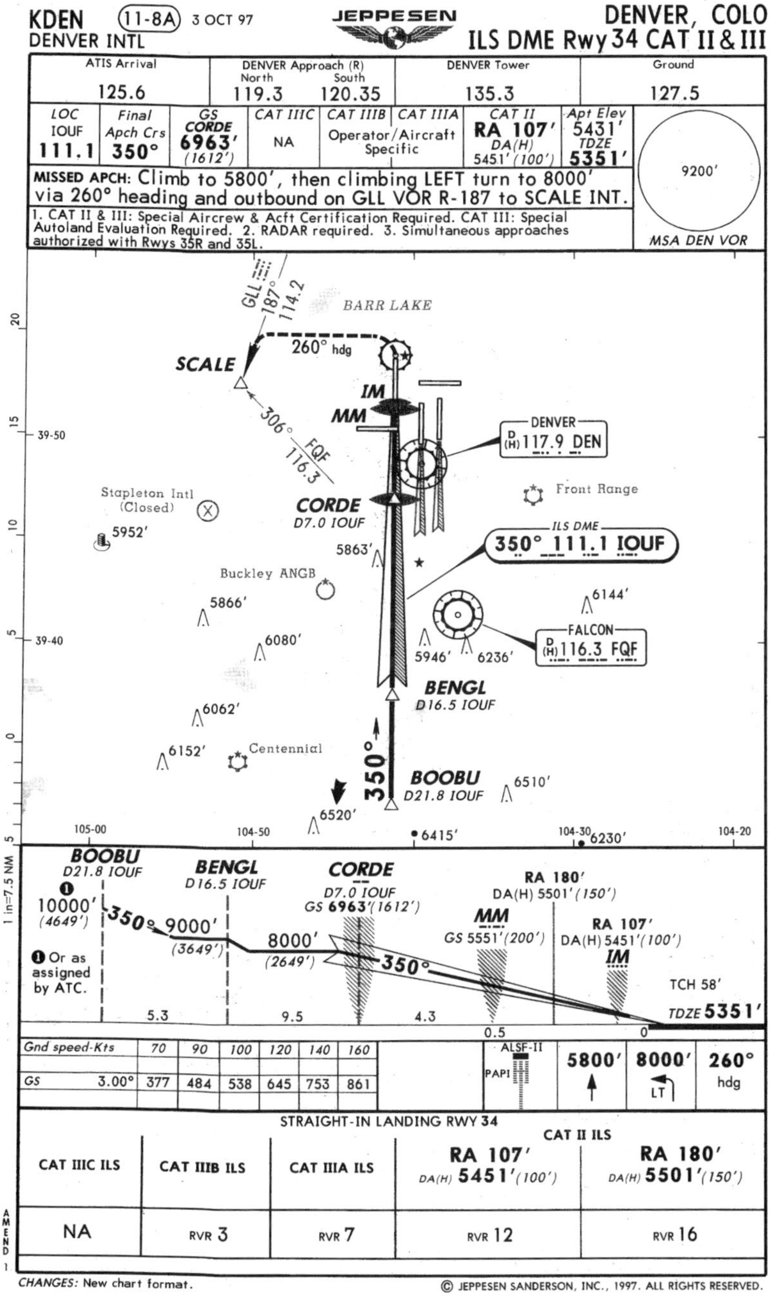

Figure 3

Denver Chart

The next line on the right has missed approach instructions. In the middle is a depiction of what the runway lights will look like for a night landing.

Below this, we see the ATIS frequency, which is selected on a VHF radio, to hear how bad it is snowing and other airport information via other frequencies for use at this airport. The middle of the chart has a depiction of the airport, including an approach to runway thirty-four which we will be flying. Note that all the towers and their heights above sea level are depicted, along with the circle in the upper left, showing 9,200 feet as the minimum safe altitude within twenty-five miles of the airport. This means airplanes should not descend below this altitude unless on a published approach such as this. Note also the location of Stapleton International Airport; it is now closed, along with the locations of Centennial Airport and Buckley Air National Guard Airport.

Below this is a depiction of descent information showing that you should be at "Boobu" intersection. This is 21.8 miles from touchdown, at 10,000 feet above sea level, which is 4,649 feet above the runway, when the descent on the ILS glide slope is started. Interestingly, all intersection names on the airway system have five letters for computer standardization, making for some creativity in names.

The notes at the bottom of the chart indicate landing information, such as visibility needed for an approach, and how low an airplane can descend. Note that for a Category II ILS, you can descend to one hundred feet if the runway visibility is 1,200 feet, but with a Category IIIC ILS, you can descend until the airplane lands itself with zero visibility. Any lawyers in the cabin?

Navigation charts for flying at higher altitudes are much more simplified, giving airway information such as course and frequency for the VOR navigation facility, along with other enroute navigation information.

Newer airplanes with 'magic stuff' in the consoles, have all this information contained in the computers, so pilots hardly ever have a need to get these charts out of their flight kits, since everything can be displayed on their television screens.

Of course, all pilots make sure their manuals are completely up-to-date before going on a flight, but I was once told a story involving a captain flying an old propeller driven airplane who waited until he was airborne to change his charts. He then put the old ones in the cockpit trash bag. After they had landed and departed the airplane, a cleaning lady thought this was important stuff and turned the contents in to the airlines chief pilot. The following day when the captain showed up for his next flight, he was reprimanded by the chief. Being a stubborn guy, he elected to do the revisions in the cockpit while flying again that day, but changed the disposal of the old charts by slightly opening the cockpit window and tossing the bag into the atmosphere. All the contents went into the air intake of the left engine. The clogged carburetor immediately stopped the engine. A lot more explaining had to be done that night concerning why the airplane had to be landed with an engine inoperative.

9

Landing

A coffee cup in my collection reads, "Flying is the second greatest thrill known to man. . . landing is the first!" Nearly all aspects of airline flying are so similar that a pilot accepts the challenge of proving to the passengers that he is the best by making the smoothest possible landing. He has no other way of demonstrating his aviation prowess, other than really squeaking the tires and coming to a smooth stop as the airplane turns off the runway. Have you ever noticed the correlation between a smooth landing and how soon the cockpit door opens for pilots to say good-bye to the passengers? If it is one that jars your bones, you will never see inside the cockpit while deplaning. The pilots will hear about it though, usually from the flight attendants complaining about the location of their pantyhose.

The actual landing of a large airplane is just like landing a small Piper Cub. Simply have the airplane centered on the runway, at the proper speed and safe altitude, then reduce the power while pulling back on the control stick in a coordinated maneuver. The airplane will then touch down very smoothly, followed by bringing the flying machine to a stop with whatever brakes or reversing is available. Should you decide to start flying lessons, remember how easy this sounds during your first landing approach.

There are two basic situations for landings, both depending on the weather. If the weather is CAVU (Clear and Visibility Unlimited), then you can see the runway many miles out. This allows the pilot to determine his glide path for landing, along with deciding when to configure the airplane. The main idea is to have the airplane in a landing configuration at the proper airspeed, combined with an altitude at which he can make a safe, visual landing on the runway centerline. If the weather is bad and Instrument Flight Rules (IFR), apply due to fog, snow, rain, or just low clouds, then the flight crew must fly an instrument approach. After the airplane is configured for an approach, speed is reduced, and glide slope is intercepted. The airplane must be configured for landing, and a more stabilized approach flown further from the runway for an IFR landing. IFR means "instrument flight rules," as opposed to "visual flight rules." A constant rate of descent on the glide slope, at the proper speed, allows the pilots to see the runway after clearing the weather, and the airplane is ready to land.

Most landings are manually flown, weather permitting, by the pilots who generally take turns with every other landing. This is opposed to an automated flight with the automatic pilot engaged. When weather minimums are below a half mile visibility, a monitored approach is flown, in which the auto-pilot flies the airplane with the co-pilot monitoring all systems while the captain looks outside for the runway. Upon spotting the runway the captain tells the co-pilot that he has control of the airplane and continues with a manual landing. If the airplane has descended to a minimum altitude with the captain not seeing the runway, the co-pilot automatically executes a missed approach, immediately pulling the nose up while applying full power, just as in a take-off. Every runway has a procedure for a missed approach that shows which heading to fly and what altitude to climb to.

Airplanes can also land automatically, as mentioned previously. This is accomplished with very sophisticated on-board electronic systems monitoring ground transmitters, sending back signals to the automatic system regarding altitude on the glide slope, and runway line-up with the centerline. Along with back-up systems, there are plenty of lights and bells to tell pilots things are not right and they need to do something differently, such as "going around" on a missed approach, unless the runway is in sight. There is a lot of technical aviation language and instrument magic in this phase of aviation, but I think this would be very boring to you, the passenger. This can be one of the "moments of stark terror," which has just interrupted "hours of boredom" on a very long flight.

Getting the airplane on the ground is just the beginning of the landing; next this heavy metal machine must be sufficiently slowed to taxi to the gate. After turning off the runway, pilots retract all the wings' additional lift devices, turn off unneeded systems, and get tower permission to taxi to the gate.

Most pilots say their greatest landing thrill is their first solo landing. Others say it was their first night landing. Some of the older guys remember coming home with battle damage. But my greatest thrill was the first time the LSO (Landing Signal Officer), said to put the hook down for my first arrested landing on a Navy carrier. Very few adventures in life will top that, except maybe the first catapult shot off the front of the same ship. A very few of us have worn Navy "Wings of Gold," which show you have landed on a boat. All other pilots simply dream of the thrill of landing on an aircraft carrier.

Shortly after World War II, when airlines from the United States started flying into Germany, it was obvious to a German tower controller that an approaching airplane was off the normal approach path for a landing. Trying to be helpful, he inquired if the crew needed any assistance, and whether this was their first landing at this airport. The reply came back from the pilot that he had been here several times before, but had never landed.

10

How's the Weather?

Weather plays an increasingly more important role in today's world. Its importance, however, is dependent upon what part of Mother Nature is most vital to you. As a farmer, the dependence is on rainfall amount; sunshine brings the sun worshippers to the beaches, and snow is important for skiers and Santa Claus. In the airline business, perhaps the two most important factors are temperature and visibility. High temperatures limit take-off weights due to the performance parameters of jet engines, while low temperatures bring out the de-icing equipment and delays associated with cold weather operations. Whatever the weather actually is outside the airplane, as a passenger you probably do not realize all the implications of weather pertaining to aviation. Originally, airplanes flew only during the day when no threatening weather was anywhere in the area. Today, passengers depend on airlines to get them to their destinations on time, regardless of the weather at the departure or arrival airports. Airplane builders have, therefore, designed newer and better all-weather operating equipment into the newer commercial airplanes. Airlines have had to spend much more money to assure that their airplanes are operating around the clock in all weather situations. Pilots are continually retrained concerning weather phenomenon affecting an airplane's operational capability. The Federal Aviation Administration monitors all these changes and decides if the operation is safe enough to meet their standards, while realizing that pilots and airlines have a much higher stake in continued safety operations in all kinds of weather.

Due to temperature changes and locations that airplanes must be able to reach in hot and freezing temperatures, most of today's jet airplanes are certified to operate at temperatures up to one hundred and twenty degrees Fahrenheit, and temperatures as low as minus fifty degrees. The limiting factor in warm temperatures is the power the jet engine can produce. Combustion of fuel and air inside an engine is much easier to achieve with cool dense air. Colder temperatures make it easier for the jet engine to produce more power due to the increased air density, but when combined with precipitation, it can also produce ice, which can have a very detrimental effect on airplanes. Airplane surfaces must be clear of ice and snow, not only to avoid disrupting lift generation by changing the shape of the wing, but also by adding weight to the airplane's surfaces. Engines must inject only pure air, with no ice, to create the correct combustion ratios with fuel. Preventing ice build-up on compressor blades where the air comes into the engine also keeps the airflow correct.

The method of keeping ice off airplane surfaces when the airplane's engines are operating is to take heated air from the engine compressor section. This air is then piped to the wing and engine leading edges to prevent ice build-up. In flight, the ideal situation is to avoid icing conditions when possible, but de-icing systems must be turned on when there is any possibility of freezing precipitation adhering to the airplane. This may mean turning the system on just for take-off, or using the hot engine air for ice protection all the way to landing. The important consideration is that less power is available for thrust production to propel the airplane forward, since "bleed air" is being taken from the engine for ice protection. You can conclude that this isn't too big a problem because colder, and more dense, atmospheric air is better for thrust production anyway.

Hot atmospheric temperatures reduce the power output of the engine and also create other problems. Hydraulic systems overheat easily, along with tires and brakes when operating at high summer temperatures. More "bleed air" from the engines is needed to cool the passenger cabin and electronic components, although the auxiliary power unit helps during ground operations.

In snow or freezing rain, airplanes have very good anti-icing systems when the engines are operating. When the engines are not yet started, ground operations require assistance from trucks equipped with de-icing fluid. This chemical, usually glycol, is mixed with water, heated, and then sprayed on the airplane. The heated fluid not only melts the ice, but sticks to the airplane's surface with a plastic-like chemical. Ice and snow removal prior to take-off is very tightly regulated by the Federal Aviation Administration. All information, including type of fluid used and start and ending times of the de-icing procedures, must all be logged into the airplane log-book. Depending on the type of fluid used and the outside temperature, the airplane may need to be sprayed more than once prior to take-off. Additionally, if the airplane does not take-off within five minutes after being de-iced, it is mandatory for one of the pilots to come into the cabin and look at the wings to see if ice has re-accumulated. The engines must be shut down for de-icing or the fluid could be ingested into the cabin's pressurization system. Hopefully, your captain will keep you well informed concerning the pink fluid being sprayed on the cabin window next to your seat.

Fortunately, a new and better system for ice removal is being developed and should be in place at a few airports in the near future. This involves taxiing the airplane through a hangar which is equipped with infrared machines in the ceiling. As the airplane passes under, it is de-iced and ready for take-off.

Rain, snow, and fog may still allow your postman to deliver the mail, but if any of these make visibility drop below published values, the pilots are not permitted to take-off or land at that airport. Visibility requirements for take-off are lower than for landing, and will change with different runways and airplanes, depending on the type of equipment installed. Newer airplanes with more "cockpit magic" can utilize Category III landing minimums, which may go as low as zero visibility, but most commercial jets are limited to a half-mile visibility for landing.

Rain and snow usually do not prevent flight operations, unless accompanied by either wind or fog. Wind is usually increased with changing weather, such as a local thunderstorm, while fog sticks around longer with no wind blowing. Airports near oceans or other bodies of water, such as LAX, SFO, SEA, and BOS, are more prone to visibility restrictions due to fog.

Cold weather visibility reductions may be combined with ice in the atmosphere, on the airplane, and all over the airport. Hot weather precipitation usually involves thunderstorms or weather fronts passing through the area, both of which can produce lightning or turbulence associated with wind shear. It may seem hard to believe, but the least dangerous of these two is lightning. Airplanes are not grounded to Earth, so if lightning does hit an airplane, it simply passes on through toward the ground. This may result in a very loud bang and flash of light but rarely any damage to the airplane.

Wind shear has caused several airplane crashes and until recently was not understood and therefore, not respected. Continental Airlines was the first airline to install wind shear detection equipment on every airplane, and now other airlines are following suit. Wind shear can be detected by radar on the ground and in the airplanes. Aural and visual warnings tell pilots of approaching wind shear. Company procedures strictly state that to avoid wind shear, a pilot must delay take-off or execute a missed approach.

Thunderstorms develop vertically due to heat rising within the cloud and bringing moisture along to the top (see Figure 4).

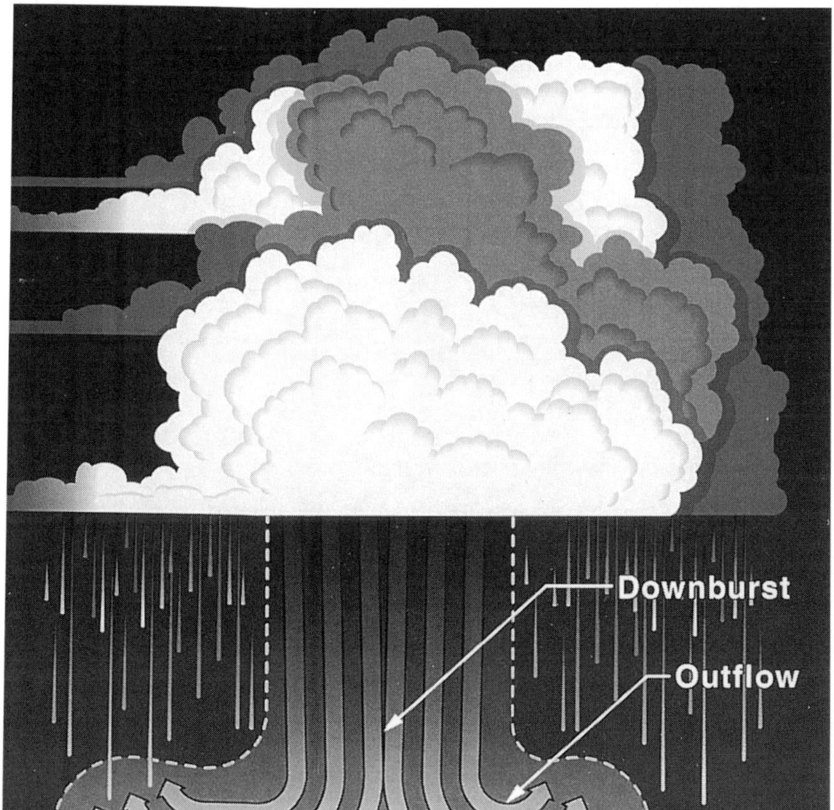

Figure 4
Weather Cell

At a point, the weight of the water and gravity overcome the heat and the water starts downward, along with the air near it. This shear between the rising and falling air causes turbulence. When the air going down hits the earth, the effect is increased, as the lifting and descending winds change from vertical to horizontal along the earth's surface. If a pilot is descending for a landing near a system producing wind shear, the power needed and attitude of the airplane will be markedly different from normal. For instance, if the wind is moving upward, the pilot will notice the airplane needs less power and the nose will be pitched downward. But stand by, because as soon as the wind shear starts, things change fast, and the pilot will need full power and the nose pointed up to avoid hitting the ground. Very rapidly changing wind conditions cause very abrupt changes in airplane performance, which must be anticipated by the pilots. Any warnings are, therefore, needed and appreciated. You can realize what these winds are like when a thunderstorm is approaching you on the ground, as the winds will increase just prior to the rain starting. Obviously airline companies and their insurance agents tell pilots to completely avoid this weather situation.

Wind conditions that are altered significantly by thunderstorms or frontal passages have a tremendous effect on airplanes flying near Earth's surface. Winds of lower speeds and constant direction also impact aviation on a regular basis. Surface winds at the airport dictate which runway will be used for take-off, as the airplane should be pointed into the direction the wind is coming from to use less runway on the take-off roll. Performance limitations prohibit taking off or landing with a tail wind over ten knots maximum. Landings are usually in the same direction as take-offs, but sometimes use a different parallel runway simply to expedite traffic flow. Winds are not always flowing directly in the path of an existing runway. This is called a cross wind, or a wind which is coming from the side of the airplane instead of directly at the nose. Crosswind operations are usually limited to twenty-five knots.

Jet streams that are shown on the TV weather broadcasts are very fast-moving wind systems which parallel the earth's lines of longitude, flowing from west to east with minor changes in direction. These high altitude winds move with the seasons, moving further north during the summer in the northern hemisphere. The velocity of winds in the "jets" can be as high as two hundred miles per hour, although around one hundred is more normal. The core of the system with the highest wind speed is about four-thousand feet thick, meaning the jet stream is like a tunnel of air moving in the atmosphere; the fastest right in the center.

You can realize that winds of this high velocity can affect the flight time of an airplane with an "indicated air speed" of 600 knots, sometimes adding hours to a flight going westbound while decreasing the flight time going eastbound. You will note that airline schedules for flights going west are longer than flights going east. Knowing the locations of jet stream winds is vitally important for flight planning. Fuel usage and associated costs of increased flight time can be lowered by knowing which path over the ground will produce tail winds of 200 knots, or how much out of the way the flight must go to avoid high head winds.

Normally, picturesque, large fluffy clouds have little effect on airplanes, unless they are towering cumulus clouds, which continue building higher due to a lifting action inside the cloud. Large cumulus clouds usually have lightning inside and are, therefore, quite visual while flying at higher altitudes. Slight changes in heading will provide a smooth flight, avoiding turbulence associated with these larger clouds. Flying through clouds when the temperatures are below freezing results in ice accumulating on airplane surfaces. Pilots must react by turning on the airplane's anti-icing systems.

All this weather information is important to the pilots and airline dispatchers. Therefore, rather than just watching the weather channel on TV, more detailed weather information must be reported on a regular basis. For many years now, each airport has sent out its individual weather conditions in what is called a sequence weather report. At the beginning of each hour, airport personnel enter the sky conditions (including clouds and visibility along with temperature, winds, and precipitation), into a coded format. These sequence reports are started with the station identifier for that airport, then the normal weather information, and end with any unusual weather activity. Pilots, therefore, have access to weather at any airport in the world, as it is reported every hour. Any significant changes can also be reported without waiting until the next hour.

Sigmets, or "significant meteorological weather reports," are also put together by meteorologists to inform pilots of area weather, such as frontal passages, approaching storms systems, freezing levels, and other useful information. These are transmitted and broadcast at regular time periods.

The most useful information is the "airport terminal information system," or ATIS, which gives all the information about the weather at a specific airport. ATIS broadcasts also include the runway in use, along with any other airport information needed for departure and arrival. ATIS is broadcast constantly and pilots must report receiving this information on their initial call to approach control. This saves the air traffic controller from telling every flight what the weather is and what runway to expect for the approach.

Until recently, the above information was put out by the Federal Aviation Administration. Many private companies are now using computers and new radar technology to provide better and more current weather information. Airlines have also updated weather-collecting equipment, eliminating the dependence on the FAA.

All the recent technology advances are going to mean fewer delays caused by Mother Nature. We are predicting weather patterns earlier and more accurately, as well as learning to better cope with what this weather produces. The advantages in safety and financial bottom lines will continue to improve, as airplane designers, pilots, and airlines better understand the effects of weather on airplanes.

11

Guys and Girls

I would guess by now that all those females who are interested in flying and have been reading this book are looking for an airport trash can, because I have referred to pilots in the male gender. Back in the 1920s, when airplanes began delivering mail, the only pilots available were those men who had survived the First World War. When airplanes were enlarged and had seats installed for passengers, the airlines decided nurses were needed on board since flying was still a new and dangerous method of travel. Therefore pilots were men, and stewardesses were nurses, who were women. As airplanes increased in size, the only thing that changed was the number of stewardesses on board each flight, but they were still all female. Women like Amelia Earhart were learning to fly airplanes, especially during the Second World War, but no women were flying as pilots for the airlines until the 1970s. Except for male service managers on international flights, the stewardesses were still all women, although not necessarily nurses, until nearly 1970.

The courts of the nation decided there was a need for more diversification, so during the seventies some women starting appearing as pilots in the front of the airplane, and more men began giving safety demonstrations and pouring beverages in the cabin. Today, just before the twenty-first century, we see an increased number of women in the cockpit, some advancing to the left seat as captains, and occasionally, an all-female flight crew. Men are approaching one third of the flight attendant totals in the cabin, and are not called stewardesses anymore.

Flight attendant training takes about six weeks, which includes training mainly in safety procedures and customer service. A new flight attendant will be on reserve, and therefore on call, for any type of airplane. They must know locations of all doors, fire extinguishers, and any other emergency equipment. Very demanding tests are given during training to make sure they understand both company and FAA standards for all airplanes they may work in. The importance of being beautiful and knowing how to apply make-up is over, as customer satisfaction is now the top priority. The day of the nurse has not disappeared entirely, as flight attendants must still deal with passengers experiencing medical problems while in flight. When you order that special drink next time, you may observe a new recruit who goes to a more experienced flight attendant to find out what goes in the mixed drink you have ordered.

Pilots are required to have a certain number of flying hours and meet other experience levels before being considered as a new hire, determined somewhat by the needs of the airline and the existing pilot availability levels. Pilots hired in the 1960s were mostly military, while now more than half have never had military flying time. Initial pilot ground school training with an airline takes about six weeks of sitting in a classroom learning FAA requirements and rules, followed by systems training for the individual airplane they will be flying. Then on to about ten sessions in simulators before actually getting into a real airplane as a co-pilot or engineer, under the very watchful eye of a check airman. A check airman is a more experienced pilot authorized to do actual training and sign off the new pilot, after demonstrating the way the company and the FAA want things done. Then the new pilot goes on his first trip with a regular captain, who will demonstrate the way things are actually done on the line. More about pilots later.

We now have women mechanics and dispatchers, while men help prepare the on-board meals along with holding the babies for parents in flight. The airlines have become so mixed you can't tell if the lady in the cockpit is a flight attendant or the captain. But do you know of any women airline CEOs? I don't . . . maybe some more changes are needed.

12

Blue Water

Bill Cosby, in his book about fatherhood, talks about his son experimenting with the blue water he found in the lavatory of an airplane. This is just one of the many humorous stories which flight crews can tell you about their experiences with passengers and lavatories. Flight crews refer to the lavatories as "blue rooms."

I was inbound to New York's Laguardia Airport, when a flight attendant reported to me that a woman had dropped her false teeth into the blue water and wanted someone to retrieve them. Naturally, the flights attendants didn't want anything to do with this and were suggesting that the pilots might know some magic trick to get the teeth back to the passenger. I told the flight attendant to use her most persuasive language in explaining to the lady that she was never going to see her teeth again. Just a minute later the flight attendant was back, saying that the lady insisted on the return of her teeth. I donned my "captain's know-all" hat to speak with the lady, and I was soon told by a frantic New Yorker that she had to have the teeth back. I remembered the weather forecast was heavy rain in New York, and informed the lady I might be able to get the teeth back to her. Upon returning to the cockpit, I instructed the second officer to tell our ground people at Laguardia that they should dump the remains of the aft blue room on the ramp if it was still raining. This is usually dumped into a truck and then into the city sewer system. After taxiing into the gate area, I made it a point to be the first one into the jetway and down the stairs, where a bewildered ground crew was sure a crazy man was testing the EPA. In the middle of a heavy rainstorm we saw the teeth come out of the airplane along with lots of other stuff. The heavy rain was washing the other stuff down the drains; the EPA would have been proud of me. We placed the teeth in a plastic bag, which I walked back up into the jetway and into the terminal where I found the lady waiting with the flight attendants. I explained where the teeth had been and that she should never put these teeth back in her mouth, and she said okay. We got our bags and headed for the front of the airport to meet our hotel van for the layover, and standing there waiting for a cab was our lady friend, teeth in mouth.

There are actually three completely different and separated water systems on the airplane, and the "blue water" is not water at all, but a chemical mixture used only in the lavatory and dumped after every flight. Sometimes if the doors or valves on the dump systems are not closed correctly, blue bombs fall from the sky after the fluid freezes at higher altitudes.

The other two systems are the potable or drinkable water, which is contained in a completely separate tank and loaded on the airplane, usually from a hose directly connected to the airport water system. These lines are then directed to all the galleys and lavatories on the airplane. Additionally, there is a drain system for dumping all fluid from the airplane except the blue water. From all the areas where fluids can be dumped, there are lines going to drains under the airplane which exhaust the fluids into the atmosphere, but these exhaust drains are heated to avoid ice build-up in colder temperatures. Flight attendants make ground personnel very unhappy when they dump hot coffee into the drain system when the airplane is on the ground.

The water system is a primary source of attention for pilots on their airplane pre-flight or inspection, especially on the first flight of the day. If you don't have water, you can't make coffee.

Airplanes must have regular maintenance checks in which the heavy, or more extensive checks, are called "D" checks. This is the time the airplane is almost completely disassembled and put back together. The containers that hold the blue water are simply filled and drained regularly, but never really inspected until they are removed during a heavy maintenance check. Maintenance personnel tell me they find the most amazing things when these tanks are completely emptied about once each year.

13

I'm From the Government. . .

The Federal Aviation Administration and the National Transportation Safety Board are the two government agencies that determine airline safety issues. Other government bureaucrats determine issues such as taxation, route authority, landing rights, legal issues, discrimination, and whether you can sue the airline over that cold turkey sandwich they said was a meal.

Supposedly the FAA tries to prevent accidents before they happen, and the National Transportation Safety Board, or NTSB, investigates them after they do. All airplanes and pilots flying today have been licensed and approved by the FAA. A pilot flying a light airplane over your house is issued a license by the FAA, and his airplane is certified by the FAA. If he crashes into your house, he is investigated by the NTSB to determine if the cause was pilot error or airplane malfunction. The ambulance-chasing attorneys take over from there.

Airplanes are licensed by the same agency as the pilots flying you to your destination. Pilots that fly little airplanes need only a student pilot's license that requires a simple medical exam, which is sent to the FAA. As more time is accumulated in the air, and hopefully more classroom hours are finished, the new pilot can take an exam to become a private pilot. Many hours later, including a lot of instrument flying time, a pilot can get a commercial license that says he is allowed to carry passengers for money in airplanes of any size. The next license granted by the FAA is the ATP (Airline Transport Pilot), which is required in order to become a captain on a commercial airplane of any size. After getting these licenses, a pilot is required to stay current with a certain amount of flying hours, along with proficiency checks given at regular intervals. A physical exam from an FAA-designated physician is required annually for a commercial license, and twice a year for captains. FAA inspectors are also allowed to fly in the cockpit at any time to check pilot proficiency, which includes inspecting licenses and currency of flight manuals. They can also check flight attendants at the same time, and probably have already reviewed the airplane on the ground with company mechanics.

When an airplane is originally built, it is put together to meet the specifications of the airline company with the blessing of the FAA. This has caused much debate over the years. Is the FAA serving the interest of the traveling public or cooperating with the airlines on airplane construction, thereby limiting construction costs and avoiding costly new items related to safety? No further comments at this time.

The FAA also controls an interesting concept called the "minimum equipment list" or MEL, for a airplane. This is a list of everything on an airplane; what equipment must be working correctly and what equipment can be inoperative and still allow the airplane to go flying. As an example, a generator creating power for the airplane's electrical system can be broken on most airplanes, because the generator on the other engine can usually provide enough electrical power. The APU (auxiliary power unit) will be operated and its generator will supply power as a substitute. Every system on the airplane can have parts not working as long as there is a back-up, or the FAA has determined the airplane can fly safely without that particular item. Minimum equipment lists specify how many hours the airplane can fly without getting the system completely repaired. An airline proposes changes in this list to the FAA, which signs off new procedures to dispatch an airplane with any given item inoperative.

All air traffic controllers are also licensed by the FAA and, indeed, are employed by the FAA. This became a national policy issue when President Reagan fired the controllers who went on strike against the government. The equipment being used to control air traffic in this country is very outdated. Air traffic controllers are attempting to overcome this handicap by keeping the system operating safely, and performing professionally when dealing with vastly overloaded traffic in our major metropolitan areas, such as New York and Los Angeles. This system is still relying on electronic equipment made thirty years ago, even though current technology exists to update safety, but is being held back by government bureaucracy at the FAA.

The National Transportation Safety Board, or NTSB, is a newer organization designed to investigate accidents of all types involving public transportation. After investigating, they are supposed to determine a cause and propose a solution to the problem. Many recent accidents, such as ValuJet in the Everglades and TWA 800 off New York, will demonstrate that the system could be improved. The NTSB has been strongly suggesting for years that all airplanes should have fire detection and suppression systems in cargo areas, but many cargo compartments in older commercial jets are still not equipped with these systems.

The government may attempt to control safety issues, but an airline with crashed airplanes all over the country will go out of business, because the traveling public will choose to patronize another airline. The profitability of airlines is greatly influenced by their safety records, which the public can read about on the front page after every accident.

The FAA has also determined that a pilot is no longer capable of flying after the age of sixty. No other profession has such discrimination based on age. After landing a jumbo jet safely one day before your sixtieth birthday, the next day you are unsafe and must never fly again. Unfortunately, one of the biggest lobby groups for this issue is the Airline Pilots Association. The majority of the members in this organization are well below sixty years old and are simply looking to gain seniority by getting into the left seat faster when the older captains hit the magic sixty and can no longer fly safely. This union obviously has a tremendous lobbying effort in Washington.

Legislation is needed to change the FAA in many ways. The air traffic control organization would operate much more efficiently as a non-government organization. A user-pay system, financed by airlines as a separate organization, would improve all aspects of air travel by updating technology as airplanes are improved. Traffic delays would decrease dramatically with newer, updated computers. Increased air travel is going to demand these changes, as passengers see increased delays at major hub cities.

Many other changes in the FAA are going to be needed as the airlines adapt to new technology and the FAA bureaucracy lags behind with old equipment and ideas.

14

How Much Does it Weigh?

So far we have discussed airplane weight, but now let's go into detail to determine how much an airplane actually weighs as you go rolling down the runway. Airplanes are actually weighed to determine the operational weight without fuel, passengers, and cargo. Additionally, the center of gravity (CG), is determined by finding out where the airplane can be balanced on a point with the nose and tail staying in the air. This point will be in the middle of the airplane's body, hopefully at an equal distance between the wing tips, and also located between the front and back of the wings. This information is determined when the airplane rolls out of the factory as a brand new airliner, and then is measured frequently thereafter, due to changes made, such as different seating arrangements or new antennas for radios, for example. Also, the FAA requires this information to be available in the cockpit and in company records.

We then have a starting point for both weight and CG. Obviously, if anything is added to the airplane, both of these figures are going to change. So, it is necessary to know how much things weigh and where they are located on the airplane.

Things like ice on the wings cannot be weighed and would also spoil lift, so must be removed before take-off.

The most important commodity on an airplane is the passenger, not only because of the human aspects, but also because they bought tickets. We know what seat you are going to be in for the CG determination, but have you ever been weighed before you got onto the airplane? When was the last time you had your baggage weighed when you checked in for a flight? Actually, average weights are used, and you automatically weigh 180 pounds in the summer and 185 pounds in the winter, due to extra clothes. Flying into Japan allows airlines to use an average weight of 160 pounds, which works unless the Sumo wrestling team is on board. Your baggage is also assigned an average weight, and the number of bags put in the airplane is counted both for weight, and to calculate how many bags are in each compartment to determine the correct CG.

Surely your three-year-old son, for whom you had to buy a ticket, doesn't weight 180 pounds. If the airplane is very heavy, you will notice the flight attendants going through the cabin looking for kids, because they can be figured as "half-weights," thereby cutting total weight significantly.

All the cargo is weighed to determine where it will be placed in the airplane for the best CG location, and also can be used to offset location of passenger baggage and the eighty-pound Black Labrador.

Fuel is usually the heaviest item, depending on how distant the destination is, what the winds are, weather at destination, take-off runway length, and the temperature at the airport. The amount of fuel needed for a given flight must meet requirements specified by the FAA. There must be enough to get the airplane to the destination, allowing for holding and diverting to an alternate airport, but taking into account winds en route. All this while still staying within runway length requirements at the existing temperature.

As all the above information is determined, we must know that the airplane can actually take-off at this weight. This is first determined by the company flight dispatcher at the operations center for the airline. By communicating with the load planners at the airport, they can determine how much cargo can be shipped, how many passengers are booked for the flight, and how much fuel to put on the airplane. The flight crew arrives just before departure and approves the figures or adds some fuel for other contingencies. Most domestic flights are well below maximum certified take-off weight, but many international flights take-off at the maximum weight allowed for existing conditions.

The maximum structural take-off weight is determined by the airplane manufacturer in cooperation with the FAA, after determining thrust available from the engines, if the airplane can fly with one engine failed, and numerous other factors. Additionally, the maximum structural landing weight is determined by knowing the landing gear limitations on a normal landing, along with power needed for an engine-out approach while maintaining a normal descent path.

Original four-engine propeller-powered airplanes could take-off up to about 150,000 pounds while the original jet airplane, the Boeing 707, had a maximum take-off weight of over 300,000 pounds (see Figure 5). The three engine Boeing 727 is limited to about 200,000 pounds while the three engine DC-10 (having much larger engines), can get up to over 500,000 pounds. Newer versions of the Boeing 747-400 have take-off weights nearing 1,000,000 pounds.

Figure 5
Boeing 707

15

Lights

When you arrive at the airport, the biggest problem is finding the correct gate and then getting on the correct airplane. The gates are all numbered and agents are supposed to check your ticket to make sure they are getting you on the airplane headed for the destination you desire. This system works most of the time, but occasionally a passenger gets on the wrong airplane. Identifying an airplane is a problem for passengers and agents around an airport, but even more of a problem for pilots when spotting other airplanes, and air traffic controllers knowing which airplane is responding to their commands on a radar scope.

Marketing plays a big part in airplane identification, since they want the company logo painted on the tail and big letters spelling out the name of the airline on the fuselage. We can recognize who owns the airplane on the ground during the day, and there are special lights that illuminate just the tail at night.

Airplane wing tips seemed the logical place to install lights when night flying started in the 1920s. Airplane wings still have a green light on the right wing tip and a red light on the left wing tip, just as on a ship. As more airplanes began flying at night, more lights were needed, so the first addition was the flashing red lights on the top and bottom of the fuselage. Originally called a rotating beacon, just as on the top of the airport control tower, it was simply a light that revolved with a mirror reflecting what appeared to be a flashing red light. This made the airplane much more visible to other pilots at a greater distance and at different altitudes. More recently, flashing strobe lights have been added to enhance night identification from even greater distances.

Landing lights are located in the wings and nose of the airplane and are used to increase visibility on the runway for landing at night. Of course, they are used for take-off also. Additionally, there are lights which light up the wing for pilots to see at night, along with lights pointed at the tail to advertise the company logo for passengers that made the mistake of flying on a rival airline. Pilots crossing near each other in the sky, have traditionally used the landing lights installed in the wings as a form of greeting and acknowledgment of each other's positions.

When you are in an airplane lined up for take-off, you will note that the airplane on the runway just cleared for take-off will turn on all the airplane lights, both for better pilot visibility and also to tell other pilots they are beginning the take-off roll.

Airport lights also have special identification according to color. As you taxi out toward the runway, the captain knows he is on the centerline of the taxiway by following green lights in the middle of the concrete, while staying between blue lights at the edge of the taxiway. All the taxiways and runways are numbered differently for identification at every airport, except that runways are numbered in the direction they are laid out on the ground. A runway that is exactly north-south would be runway 36/18, while a east-west runway would be 9/27, thus testing your high school geometry lessons. Runway lights are light yellow on the sides and white in the middle, changing to red for the last 3,000 feet. This is to tell pilots they need to get into the air for take-off or, if landing, it is time to really increase braking before running out of concrete.

Air traffic controllers have a special way of identifying airplanes, since they obviously cannot see them. During World War II, shortly after radar was invented, another black box was designed which displayed a special code for each airplane, rather than just a lot of blips on a scope in which the controller had no way to tell which airplane was which. A four-digit code was assigned to an airplane, and from military terminology, the pilot was told to "squawk" a certain code or dial four numbers into the box in the cockpit. The ground radar then displayed not only the radar return blip, but also the number identifying the individual airplane.

As electronics progressed, the controller's scope readout also displayed speed, altitude, and course information as he followed the airplane on a airway or highway in the sky. Separation from other airplanes became much more efficient and more airplanes could use the airways to navigate to more cities, therefore, carrying more passengers in the air. Today, more changes have made the system more efficient as we get closer to depending on satellites, not only for navigation but also for separation from other airplanes.

Lights of other airplanes are visible for only a few miles in the night sky. As a substitute for actually seeing other airplanes, a recent invention is TCAS (Traffic Control and Separation), which is installed in all commercial passenger airliners. The radar scope in the cockpit previously displayed weather information, but could also paint "blips" where other airplanes were. Now TCAS also provides information about the other airplane's altitude and course. If the other airplane becomes a threat by closing on your airplane, the two computers in both planes talk to each other and suggest altitude change information to eliminate the conflict. Any pilot will tell you that TCAS is one of the greatest inventions in modern aviation. Some FAA controllers are arguing that pilots are taking control from them by using this system for aircraft separation, and the debate will continue for a few years. Are they air traffic controllers or air traffic coordinators?

16

Communications

Communication in airplanes is becoming much more advanced as technology continues to improve with new systems. There are many ways to communicate in and from modern airplanes. You, as a passenger, are aware that the captain and the flight attendants can use the public address system on the airplane to give you directions and also keep you informed about the flight. This system also is used for the cockpit crew to warn the fight attendants that they have been cleared for take-off and landing, so they can get into their seats with seat belts fastened. In case the passenger address system fails, there are hand-held battery-powered microphones that can be used.

Airplanes also have telephones built into the seat so passengers can call anybody at a somewhat higher charge than normal. The cellular phone cannot be used since there is some evidence of interference with navigational radios in the cockpit. Maybe the airline just wants you to use the more expensive phones they put on the airplane.

Communications from the flight crew to ground facilities are obviously very important. The primary system for this is

VHF (Very High Frequency) radios, of which there are usually two installed. The number one radio is used for communications with ATC for monitoring position both in flight and on the ground. The second VHF radio is a back-up, but is also used for listening to weather reports and communicating with the airline personnel. Weather at individual airports is monitored by the ATIS (Airport Terminal Information Service), which is broadcast continually, giving pilots information such as the active runway, visibility, wind, temperature, and any other information which would have an effect on the landing or take-off. This radio is also used to tell the airline of special needs, like more ice for the drinks or a need to dump the blue water. The company personnel tell inbound pilots which gate to use, along with other pertinent information like an airplane change on a scheduled through flight.

"SelCal," or selective calling, is a system whereby the radio not being used is always tuned to a known frequency. It can be called by ground personnel while the airplane is in flight to pass on information to the crew, such as, "the weather at the destination just went below landing minimums so get ready to go elsewhere." Another use of the extra radio, just recently started, is MedLink. As you can tell from the name, this is a method for flight crews to communicate with medical personnel when a passenger gets sick. This has turned into a terrific system. Specialty doctors on the ground can give instructions to a medically trained passenger on the airplane regarding a passenger that needs medical attention, or simply advise the crew if an earlier landing is needed to possibly save a life.

A relatively new system, ACARS, is a system which automatically transmits and receives information about a flight, such as take-off and landing times. This system also transmits to the airline company the time the first cabin door is opened for arrival or when the last door is closed for departure. On-time arrival information is taken from this system for most airlines. On more advanced ACARS systems, more information, such as passenger cabin seating, gate changes, and other important flight data can be automatically transmitted to or from the flight crew.

VHF radios have a limited range, so on oceanic trips the ground to air communicating is done with high frequency radios which have a longer range, but also have lesser clarity, which makes long-distance communications somewhat outdated. You can telephone someone in another country but airplanes still use outdated long distance communications.

Airplanes and ground stations always monitor "guard channel," which is the priority frequency for any airplane which has an emergency. By declaring your identification, position, and type of emergency, immediate help will be on the way, and other pilots in the area or ground stations are all listening in and ready to help.

Transponders are radios into which a pilot selects an assigned four digit code by which ATC can know specifically which radar blip on their screen is a specific airplane. Pilots can change this code to tell controllers of a problem. As an example, when pilots feel they have an emergency situation, this code is changed to "7700," which lights up the controller's screen to make him aware of an airplane with an emergency. There are also special codes to alert controllers of lost communications and hijacking attempts.

Air traffic controllers also have land lines to talk to each other without interfering with the radios they use in talking with pilots. When transferring a flight into the next sector or coordinating a pilot's request for direct routing, the controllers talk to each other to resolve the flight paths of many airplanes. Of course, pilots that don't get an immediate answer from a controller are then told they were on the land line, so pilots sometimes use the same excuse when they miss the first call from a controller while dining on a crew meal.

Additionally, pilots communicate with the ground guys driving the tug that is pushing you out of the terminal area. There are also plug-in jacks at several points all around the airplane for mechanics to talk to the crew about a problem in that area of the plane.

Perhaps the biggest area that needs change now is long distance oceanic flights, which rely on HF (High Frequency), radios, which are terrible for communications. There is a need for a new system which can use satellite relay to communicate long distances. Additionally, airplanes flying over the oceans are not in radar contact, therefore a positive radar control system, like the one used for domestic flights, does not exist. Technology for this is available, but the government agencies do not feel this is needed yet, even after a few close calls in the North Atlantic.

All communications between pilots and ground-based air traffic controllers are supposed to be in English. This is true not only in the United States, but in every other nation. Although most foreign controllers observe this rule, some nationalities have too much pride when communicating with their own countrymen, especially close to Paris or Montreal.

17

Warnings

Modern airplane systems have become so complex that continuous monitoring of all the different parts of all the different systems is nearly impossible. Additionally, flights are becoming longer, introducing fatigue problems, along with much more traffic in the sky, which means the pilot must be more aware of the situations outside the airplane. Built into the airplane systems are methods of telling pilots when part of the system needs attention, indicating problems such as low pressure or a failing component. All of these warning systems must be tested before each flight. If you are seated near the cockpit, while on the ground, you can usually hear a lot of bells, horns, and voices which are being tested to be sure all systems are operative for the upcoming flight.

GPWS (Ground Proximity Warning System), has been publicized a lot recently due to accidents. Basically, this system measures the distance from the airplane to the ground, and when parameters are exceeded for closure rates, the pilots are warned with both lights and voices. If, after take-off, the airplane starts descending while still below 1,000 feet, the pilots are told to climb. If at cruise altitude, and a collision with high terrain is possible, the GPWS warns of terrain proximity. On landing, if the plane is performing a high descent rate or getting too low without the landing gear extended, the system will also warn the pilot that the landing approach needs to be changed. New technology is now utilizing satellites to give more information to pilots as to how close the ground below actually is.

TCAS (Traffic Collision Avoidance System), tells pilots of the locations of other airplanes. If they get too close, it will advise whether to climb or descend to eliminate the threat, and will also tell them when they are clear of danger. This is a fantastic system that will soon advise turns as well as climbs and descents.

Wind shear is a weather phenomenon associated with thunderstorms. If a thunderstorm is near the airport, the best procedure is to enter a holding pattern for a while because the tower people have warned of a thunderstorm in the area. But there are times that wind shear can be caused by other invisible weather situations that the airport people do not know about. Now pilots have a wind shear warning system in the cockpit. A loud voice saying, "wind shear," warns pilots that the possibility of a shear exists, therefore, it is best to get out of town immediately and talk about coming back a few minutes later when more weather information is available.

Most of the other warnings are for airplane systems, such as a warning horn if take-off power is applied by advancing the throttles forward, and the airplane is not properly configured for take-off. If the flaps are not in the take-off range, the elevator on the tail is not correctly trimmed, or the parking brake is still set, an intermittent horn will tell the pilots to abort the take-off until the situation is corrected.

If the air the passengers are breathing in the cabin gets too thin due to a pressurization system problem, and if the cabin altitude level reaches a certain point, a horn and light will tell the pilots to reset the pressure in the cabin. If this does not solve the problem because of a cabin door being open, for example, you will soon see the yellow masks coming out of the ceiling for passengers to use.

A clacking sound, along with a vibration of the control column, means the airplane is flying too slow and is getting close to the stall speed, meaning not enough air is going over the wings to generate lift. Pilots immediately apply full power while pushing the nose over, if the airplane is in the air. But on the ground, pilots can entertain young passengers with this warning system by telling them we have a machine gun in the nose—that is exactly what it sounds like.

A loud bell, accompanied by red warning lights, will warn the pilots of a fire in an engine or the APU. It will identify which engine has a problem and what handles to pull to discharge freon into the hot area. Each engine has two systems for back-up purposes, and these are checked if only one system indicates a fire prior to shutting down the engine and proceeding with fire extinguishing procedures.

Perhaps the most embarrassing horn and light for a pilot would be the landing gear warnings when close to the ground, indicating the crew is getting ready to land and the wheels are still up in the fuselage. This situation can lead to a "landing at a lower altitude," and also takes a lot of power to taxi to the gate. A very noisy cockpit, both from horns and "why didn't you tell me?," from the other pilot, usually prevents an actual wheels-up landing.

Pilots also get warnings on a chime system from the flight attendants indicating the cabin temperature or other cabin situations. If the situation isn't corrected, it is usually followed by a knock on the cockpit door saying if it doesn't get warmer back here very soon, you can kiss-off that crew meal.

A loud clacking sound, usually on the last trip of the day, can be heard from the cockpit, which means the airplane is going faster than design limit speeds, which on most commercial jets is nearly 90% of the speed of sound. As a passenger, when you hear this you will also notice a reduction in engine noise as the throttles are pulled back to reduce power.

Of course, the biggest warning light for a pilot is lightning at twelve o'clock, or straight ahead of the airplane, which will be followed by a call to ATC to get permission to turn immediately to avoid bad weather.

Every different airplane system has many dials and lights, but also has a built-in means of warning pilots of a problem in that system. Every door on the airplane has a light and all hydraulic systems can tell of low pressures. If the auto-pilot is disconnected, the pilots are warned. Descending below landing minimums turns on lights and voices, and trying to land with the landing gear not down causes lots of red lights, horns, and other "verbal warnings."

Newer airplanes not only give pilots lights and horns, but also present the system diagram with the problem area on a CRT screen, which looks just like your television. The information will tell them what to do and what will be inoperative after they do it. This is another example of how safety is playing the primary role as new technology becomes available to make the pilot's job easier and your flight more comfortable. If you are on the ground with some extra time in the airplane, most pilots will be glad to demonstrate some of these systems to you if you ask the flight attendants to ask the captain for a short cockpit tour. Cockpit doors must be locked when a United States airliner is moving, but this is not so for foreign airlines.

Maybe someday we will have a new warning system which tells the flight attendants the pilot's coffee cup is empty.

18

Black Box

Following an airplane accident, all you hear on the television is if they found the black box and if it is on the way to Washington to be analyzed by the experts at the NTSB. What is this magical black box?

First, it isn't black but painted a very bright orange to make it easier to see, and secondly, there are two boxes. One box records airplane information and the other records audible information from the cockpit. The recorders are physically located in the back of the airplane, because this is the section which is the most likely to avoid damage in an accident. Additionally, a transmitter is attached to the box which transmits a continual signal on emergency frequency once activated by an accident or submersion in water. This makes finding the box easier when an airplane "lands" in water.

Most of us want to hear details of the conversation from the cockpit to see if they knew what they were doing by discussing the pending emergency. This can be determined by a microphone in the cockpit that records all conversation, along with everything that is said on all radios and intercom in the airplane. This continuous tape is always recording over itself, so that it only contains the previous thirty minutes of cockpit conversation. Also, this tape can be erased if the airplane is on the ground, with the parking brake set, and the pilot depresses an erase button for seven seconds. This erasing procedure actually becomes common practice if company officials decide to monitor these conversations to check on crew compliance with checklist usage and other operational situations.

Original flight recorders on airplanes simply recorded five parameters of airplane flight performance, such as airspeed and altitude. New technology and recorders have made possible the recording and transmitting of all aspects of the flight, such as control positions, glide slope relative position, optimum speed and altitude, and much more information. This is valuable not only to the NTSB in an accident investigation, but to the FAA in regulation enforcement, and the airline company in monitoring pilot performance. This information is not only recorded, but continually transmitted to operations departments of the companies that employ the pilots. This is causing a lot of unhappiness to pilots in dealing with their employers, but does allow airlines to determine on a continual basis which pilots are not flying airplanes in the style dictated by the company and the FAA.

These recorders do not record anything that happens in the passenger cabin which could be of value, such as conversations with unruly passengers, terrorist hijacking situations, or what actually happens in an emergency evacuation. The original concept for recording flight information was strictly accident investigation, but now the recorders are used for flight crew compliance and operations procedures as well. A pilot can disable the recorders by simply pulling a cockpit circuit breaker for each recorder, should an incident arise for which a recording may prove to be dangerous. Of course, in an accident there is not time to pull any circuit breakers. Cockpit conversation is not supposed to be public record, but only used in an accident investigation.

The bottom line is that a "black box" records voices and airplane information that is valuable in investigating a mishap, but the government agencies and airline management are sometimes mishandling this information, in attempting to point the finger at pilot error.

19

What About Fire?

There are controlled fires in the burner cans of each engine, which is where the fuel and air are mixed to produce combustion, thus pushing the engine forward. This is the only fire allowed on an airplane. Throughout the rest of the airplane there are fire detectors installed in any area where a fire could possibly develop, and smoke detectors in many other areas, including the lavatories, where passengers may attempt an undetected cigarette. Uncontrolled fires of any kind on an airplane are bad news and must be controlled and extinguished immediately.

The controlled fire in the engine can become uncontrolled if a mechanical failure occurs, allowing the fire to expand beyond the combustion chambers. Unintentional fire in the engine area can be caused by a fuel leak, a broken hot air line, hydraulic fluid escaping from a cracked line, or an electrical short circuit. Most jet engines have duplicate loops between the engine and the cowling, which detect any overheat situation and warn the flight crews immediately. The pilots then can determine the source of the problem after reducing thrust from that engine. If the fire warning stops, the crew can elect to let the engine operate at a lower power setting. If the warning continues, the engine is shut down. If the fire warning continues with the engine not operating, the crew can inject freon into the engine, which will extinguish any remaining fire.

Recently, fires in the cargo areas have been of great concern to the public, due to the accident in the Florida Everglades. Smoke detectors are installed in cargo areas in the bottom of the fuselage in most airplanes. However, very few of the older jet airplanes have extinguishers built into cargo areas below the passenger cabin. The theory was that by closing doors, preventing ventilation from the main cabin above, the crew could deprive the area of oxygen and the fire would go out or be prevented. The FAA designates these areas as class "D" compartments; the idea is that they will smother the fire and its source due to lack of oxygen. More modern airplanes have better detection systems, including back-up systems and extinguishing methods in all cargo areas. Fido and Boots prefer the newer systems for their air travel.

Fires in the main cabin area have almost been eliminated with the cigarette ban imposed by most airlines. Generally, the other area for concern is the galleys, which are a source only when a flight attendant puts paper in the oven or there is a minor electrical problem compounded by liquid spills. These are eliminated by removing the source of the problem, but if they get out of control, fire extinguishers are located throughout the cabin and in the cockpit. Emergency breathing equipment is also conveniently located for flight crew use near each galley and in the cockpit.

Outside the airplane the main problem area for smoke and fire is the landing gear. Overheating brakes cause smoke, but hardly ever fire, because the tires have overheat relief valves which let the air out of the tires at a certain temperature, thus preventing explosions. Broken hydraulic lines which become heated can get quite smoky, but usually dissipates as the fluid disappears. Any smoke outside the airplane will interrupt the dominos game at the airport fire house, resulting in the quick arrival of fire equipment to aid in the smoke elimination.

But what about the fire from a lightning strike while flying near a thunderstorm? There will be no damage to the airplane you are flying in, since it is not grounded. The small extensions coming off the trailing edge on the outside of the wing area are static discharge wicks, which are designed to let the lighting leave the airplane after entering at the nose. A lightning strike is very noisy, but does little damage to an airplane and will not cause a fire. You may see a fireball, however. This is simply static electricity looking for an escape path from the airplane as the lightning travels to the ground. Following any lightning strike, maintenance people are required to do a complete exterior check of the airplane, along with the seats in the cockpit, before it goes flying again.

On a hot day while you are boarding an airplane, you may notice what appears to be smoke coming from the cabin ventilation system. Do not be alarmed, as this is not smoke, but simply condensed air resulting from the high temperature and humidity.

Again, new technology has almost eliminated fires from airplanes, and extinguishing systems are nearly perfect. If an overheat problem develops, you will probably never know it, as the flight crew will have the problem corrected without causing alarm to the passengers. You may notice the smell of well-done cookies coming from the galley area, or a flight attendant reprimanding a passenger for smoking in the lavatory. Both are good!

20

Surviving a Crash

Airplanes today fly millions of miles, with thousands of passengers on board, to many different destinations, and almost all of these airplanes are retired and cut up for scrap metal after being stripped of all the reusable equipment, like seats, radios, instruments, generators, engines, tires, and anything else that can be used in another airplane that is still flying. However, as we all know, occasionally an airplane does not make it to its final destination and park at the arrival gate the way it was planned. Most of these incidents are very small, involving a minor mechanical problem which is quickly repaired, and the airplane is back in service. But the television news channels are filled with all the grim details when an airliner does hit the ground at a location other than a runway.

As a passenger, you have no control over whether or not you survive an accident, with the possible exception of how you react after it happens. Of course, it is very important that you have your seat belt fastened correctly over your lap and all your carry-on items correctly stored. Your best reaction after an accident would be to get out of the airplane as quickly as possible, at the exit that is nearest to your seat. Flight attendants are trained to keep you from exiting through a door involved in a more dangerous situation, such as a fire. This may not be easy if it is night, or if the cabin has filled with smoke. Lights installed in the floor just for this situation will guide you to an exit. You may need to be crawling on the floor rather than walking, to avoid smoke and heat at higher levels in the cabin.

If you are the first one to arrive at an exit, you will probably see it has been opened by a crew member. Remember that flight attendant seats are at exit doors for this reason. If not, there will be a handle to open a cabin door, which could happen very fast, or a handle you can grasp to manually open an over-wing exit. After removing an over-wing door, toss it outside so it is not in the way of other exiting passengers.

After exiting the airplane via a cabin door, you will see a slide has inflated. You should jump on it with your feet out and slide down on your back until you reach the ground, where you should get away from the airplane quickly. On an over-wing exit, you may have to walk on the wing toward the tail of the airplane and slide down the flaps. When on the ground, get away from the airplane.

A very good habit to develop is noting where the nearest exit is, relative to your seat, when you initially get on an airplane, and do not necessarily attempt to go to the door where you entered.

Fire is the biggest danger to passengers on a crashed airplane, therefore, you want get away from the airplane as quickly as possible.

Many airports are near oceans or rivers, where the chance of surviving a "landing" in water is much better than landing on ground not part of the airport. Airplanes will float for a while if no extensive structural damage has occurred, and water is softer than dirt. There is a life vest below your seat, your seat cushion will float, and the slide at each door will also float. If you are in a situation where you will not be rescued immediately, there is a complete package of survival gear on the airplane and the crew members are all well-trained in its use. Rescue people will arrive at an accident site very quickly at an airport location, and soon thereafter at other locations. Follow the instructions of crew members and help other passengers if you are able. If time permitted, pilots have notified air traffic controllers that a problem exists and that they should start the rescue procedure.

Airlines have recently modified the post-accident assistance rendered to passengers and relatives. Obviously, the press will want to get your thoughts for the nightly news, but note that no crew members will be talking with the news media. Your next dangerous encounter will be with attorneys wanting to "help" you start taking legal action against the airline.

I need to end this chapter by saying that after forty years of flying I have not been involved in even a minor incident. Travel on the airlines of the United States is the safest transportation ever, and getting safer every day.

21

True Stories

The airline story which I tell the most started when I was fourteen, being inducted into the church, in the mid-50s. As part of our reward for finishing the entire course, we were taken to Washington, D.C. to see the sights, one of which was Washington National Airport. Some of you may remember the days before airport security, when you could almost walk out to the airplanes. For a young guy who loved airplanes and wanted to become a pilot, the idea of seeing the planes at close range was terrific. The only obstacle to getting outside on the observation deck was a turn-stile, which required the deposit of a dime for the arm to move. Four of us decided to keep our dimes and simply duck under the arm for a free tour of the DC-6s and Constellations.

A week later, after returning home, Pastor Jack called the four of us together and told us that he knew we had snuck out onto the viewing area without paying the dime. Our punishment for someone in the group squealing on us, was to write a letter of apology to the director of the airport, including a dime in each letter.

About thirty years later, I had the opportunity to be the captain of a 727 flying into Washington National Airport (see Figure 6). After landing and taxiing to the gate, I deplaned and went inside the terminal, looked up Pastor Jack's telephone number, and called him. I explained I had again gotten onto the ramp at Washington National Airport without paying. I heard him laughing for at least five minutes. Two weeks later I received a letter saying he thought the story was so good he had used it in his sermon on Sunday. Pastor Jack Sammel will receive a copy of this book, for which I shall charge him ten cents.

Some stories of the past, which were probably true at one time, tend to lose their credibility over time, but they still makes good "true" stories. On one of my first airline trips a story of this type was related to me.

Supposedly, a new co-pilot on his first trip with a salty old captain was bouncing across a very hot Texas landscape at 4,000 feet in a two engine propeller-driven airplane. After the fourth take-off, the captain reached behind his seat and pulled out two box lunches and handed one to the co-pilot. The co-pilot knew it was very hot in the cockpit, as well as outside, and felt the stewardess would shortly bring a drink for the two working guys. Taking things into his own hands, the captain reached into his flight bag and pulled out two bottles of beer, uncapped both of them, and handed one to the new co-pilot. Being very surprised, the co-pilot volunteered to the captain that they couldn't drink while they were flying. The old Texan captain looked at the new guy and said, "Hell boy, this ain't drinkin', this is beer!"

Some of the most memorable parts of my career as an airline pilot, and as an ardent football fan, would be flying as a captain on charter fights for National Football League teams. Occasionally, we would be able to stay overnight where the team was playing and then take them back home after the game. This usually meant some smooth talking to the team's public relations man for tickets to the game. The success of this

Figure 6
Boeing Model 747-400

effort was usually directly related to the quality of the flight attendants on board.

I happened to fly the Raiders several times and found the head coach enjoyed a cockpit visit. During one visit, while cruising at altitude, we were approaching a rather high cloud, so I simply moved the heading selector on the auto pilot and the airplane followed with a nice easy turn until we were past the cloud and back to the previous heading. I remember the coach saying he would be right back. He soon returned with nearly all of the coaching staff, to whom he wanted to demonstrate this turning technique. When the coach was ready, I again turned the airplane and turned back to the assigned heading. I then commented that it would be great to get the entire team to turn like that on a football field. I received a "highly unlikely" laugh.

NFL teams eat very well on these charter flights. The amount and quality of the food put on board after a game is amazing. A lot of the huge steaks go uneaten, as players simply fall asleep right after getting on the airplane for the flight home. One flight, taking a team home after losing badly at Mile High Stadium, was boarded with so many huge filets that each player could have had at least two. The coach was so unhappy with the team's performance that he dictated that not one player on the team was going to eat anything. The flight attendants, realizing the steaks would be put in the trash, offered the pilots the steaks they didn't want. I replied my Labrador might like one or two, so I received a bag with a dozen steaks for the ride home. My Lab was very happy to be dining on the losing team's steaks.

22

Latitude and Longitude

From the time living things first started moving around on Planet Earth, they had to know where they were, or more accurately, where home was. Creatures departed their homes in a quest for food, or other things they felt important, but always knew where home was for their return. Therefore, all navigational information was based on how far, and in which direction, they had traveled from home. Distance was usually measured in time, which was generally determined by the location of the sun. Direction was usually determined by reference to a navigational aid, such as a mountain, river, or other prominent landmark. This system works great for animals that stay close to their home or know migratory routes.

When man and woman got started here on Earth, everything changed. The original humans needed a cave to live in, but had to go out to search for food, firewood, water, and, of course, the search for a mate to share their cave. Distance and direction became more important as they began wandering further from home and going in different directions. Having the ability to actually think led man in search of other things which would make life easier, but that sometimes meant going for longer periods and in new directions. Of course, at this time, there were no compasses or clocks, so navigation information was based on remembering where one had been, how many days it took to get there, and when it was time to head for home again.

A few thousand years later, the population of the planet was dramatically increasing, and man was searching for new places to live, hunt, trade, or move to for a better lifestyle. Simple migratory movements, where there was no need to return home, eliminated the need for navigational information, because no one cared where they were or where they had been. However, any time a man went exploring and left a good-looking lady to take care of the kids at home, he usually found a way to mark the trail so he could get back to a good comfortable bed and home cooking. This meant better systems of navigating, and knowing one's location on the earth became even more important.

Land movements of man were generally pretty easy as established trade routes came into being, allowing people to travel large distances and then return home again. Navigation started getting more complicated when flotation devices, capable of drifting over water, could travel away from the sight of land. Relying on the sun, or learning the stars for night travel, meant boats could start taking man away from land masses for the first time. Other information, such as ocean currents, tidal systems, and bird migrations, assisted the early marine navigators.

As the Crusades were changing the history and geography of the large Europe-Asia land mass forever, Norsemen from Scandinavia were also looking for places to expand on the oceans, after conquests ranging from England to Russia. Having built the best boats for ocean travel, the Vikings headed west over the ocean, where no man had been before. They could do this without navigation systems, since they simply maintained westerly direction for about six days and nights and arrived in Iceland, passing other familiar islands en route. Another week heading westerly and they arrived in Greenland.

Then just an additional week of travel brought them to what is Newfoundland in Canada. All this travel was a straight line headed due west, which could easily be determined by the sun during the day or the stars at night. Coming back home was just as easy and probably faster, due to westerly currents and winds in the North Atlantic.

Navigation, even over the oceans, was very basic until European powers decided they needed to expand economically. Investments in colonies in other parts of the world enabled them to search for products not available in Europe, and labor from Africa. England and Spain fought a large sea battle to determine supremacy of the oceans. England won and started looking for the westward route to China, but Spain, Portugal, and the Dutch also joined the chase, all in search of new empires for trade and conquest.

The problem was determining position on an open ocean with no land in sight for many days. Previously, man could say how far he was from home and which direction he had to go to get back. Now with long ocean voyages, navigators had to determine position in a new, more accurate way. Two new terms, latitude and longitude, were invented for sailors to determine where on the earth's surface they were in relation to the equator and the poles, but also from Europe, so they could return home or at least know how far away they were. These theories actually started many centuries earlier, with men going against the teaching of the church as to what was the center of the world. Some were soon convinced that the sun was actually the center of the universe and this revolving planet just a small sphere, where man just happened to be learning to navigate by learning the concepts of latitude and longitude.

To understand our position on this sphere called Earth, just realize that the north and south poles are at the top and bottom with the equator running around the middle. You could divide this distance between the poles and the equator into equal segments called degrees by realizing a circle has 360 degrees, therefore, one fourth of that would be ninety degrees. Sailors could determine their latitude north of the equator by locating the North Star, which is always located directly north, and simply measure how far it was above the horizon to determine their latitude. If you are at the North Pole, the North Star is directly above you. Halfway between the equator and the North Pole, the North Star is elevated at an angle of forty-five degrees, therefore you are at forty-five degrees north latitude. At the equator the North Star would appear right on the horizon, if you could see it. Therefore, determining latitude in the northern hemisphere was easy. Southern hemisphere navigators used a constellation called the Southern Cross to get the same results when south of the equator. This groups of stars now appears on the flag of Australia.

What was needed now was a method to determine their distance from home, because they knew how far they were from the equator. They knew how far north or south, but not east or west. China was where everyone wanted to go, but no one knew how far away it was, or whether the best way to go was west or east. Another part of the problem was that no one knew how far South Africa extended, or if it actually was a land bridge past the south pole and beyond. The early Portuguese figured this out by getting around the Cape of Good Hope and heading back northeast to India and China. Christopher Columbus badgered English, Spanish, Portuguese and Italian bankers, as well as kings, to get boats which he planned to point west, what he knew was the shortest route to China. We all know the history from this point, but understand that European monarchs were offering huge financial rewards for the person who developed a way to determine longitude.

Clocks for determining time were originally invented for monks, to remind them when they had to pray in the monastery. But developing a clock that could still operate on small ships on a large ocean, with continual movement from pitching decks, was a problem. We knew how many hours there were in a day, that the earth was round, and that it revolved so that the sun "rose" at the same time each day. The rest of the problem was solved by free enterprise.

Chronometers were very accurate clocks, guarded with intense security from the other crew members and other nations, so that only the captain had the key to check what time it was in London, or more specifically, Greenwich. Greenwich Mean Time, or GMT, is the same as Zulu time in aviation talk.

Navigators used chronometers to find longitude by determining how many hours they were from England. By sighting the sun directly overhead, they could easily determine how far west of England they were in miles, if they knew how big the earth was, but no one knew for sure. In 1492, Columbus was sure that he was in China when he was only halfway—in the Caribbean.

Navigation didn't change much for a few hundred years until airplanes started leaving land areas, just as boats had taken man away from the shorelines many years before. Everyone now understood latitude and longitude, but needed to determine position in a better way than looking out a cockpit window for the North Star.

As discussed in Chapter Seven, long-range oceanic navigation systems were developed, but much more accuracy was needed. More recently, Inertial Navigation Systems were developed to give pilots on long-range jets not only their exact position over the earth, but also speed and course information. But the military forces of our country needed even more accurate systems for submarines, bombers, and other secret systems being developed to defend us from the Russian enemy.

Satellites, with fantastic cameras, were going into orbit to spy on other countries. Why not use this same technology to help navigators determine position? Global Positioning Satellites were soon orbiting the Earth, relaying exact latitude and longitude information to pilots as well as submariners, tank commanders, infantrymen, and even Cadillac owners with big bucks, so they could still keep their portable GPS units on their boats at the marina.

GPS navigation is so accurate it will soon be used to warn pilots of other approaching airplanes. Satellite information can also tell a pilot the airplane's altitude. We have come a long way since Hagar and the Vikings headed west to beat Columbus to America.

23

Airline Finances . . . Big Dollars

Large airlines consume large amounts of cash to get started and to continue operating, because of the large investments required in aircraft, facilities, and training. Even for a new airline to begin operating with three used 737s, it would take tens of millions of dollars to buy or lease the planes, acquire or lease gate space at airports, and train their pilots, flight attendants, and agents. As airlines grow, they usually are able to generate the cash from customers, and as they develop, they can do all these expensive functions more economically. The huge sums of money required are why we generally see a few large airlines and very few new airlines trying to get started.

Airlines have cost structures which are generally fixed. That means that it will cost almost as much to fly an aircraft on a given trip with one passenger, as it will cost to fly 200 passengers on the same trip. Looking at some of the large expenses to run an airline explains this. In descending order, the largest expense items for an airline are: salaries and benefits (including management, ground personnel, and flight crews); fuel; equipment costs (including aircraft and facilities); and travel agent commissions. For each dollar of revenue, a typical airline will spend about thirty cents on salaries, twelve cents on fuel, eleven cents on equipment, eight cents on agent commissions, and thirty-four cents going to other costs, such as food and beverages, maintenance materials, and airport landing fees. These will be about the same, regardless of the number of passengers on the flight. There is also very little an airline can do to affect these costs; union contracts determine salaries, the aircraft and gates are already purchased, fuel price is what it is, and landing fees are determined by the weight of the aircraft when it is full.

With most of the costs already determined when a flight is scheduled, airlines do all they can to manage revenues, and profitability on each passenger mile is measured in hundredths of cents. New jet airplanes cost from about twenty-five million dollars for a small regional jet, to as much as one hundred ninety million for a new 747 (see Figure 7). These jets usually last about twenty years, but even with that length of time, it will take thousands of ticket sales in each seat to justify the aircraft purchase. The fixed cost structure is also why airlines tend to either make very good profits or have very large losses.

With the costs already spent, airlines often try to fill the last few seats on any flight at almost any price. Once the aircraft has departed, the airline has lost any opportunity for additional sales dollars. Airplane seats are like theater seats; they cannot be stored. So, airlines try aggressive marketing tactics to fill as many seats as possible, every single day.

Key operational terms for airlines are revenue passenger miles (RPMs), available seat miles (ASMs), and load factor. One revenue passenger mile is generated when one passenger flies one mile. If 100 passengers are on a 1,000 mile flight, then 100,000 RPMs will be generated. Multiply this for all the airline's flights and you will have total RPMs. RPMs usually correlate to an airline's size, revenues, and passengers. Available seat miles are similar, except you multiply the seats available in an aircraft (either filled or empty), to measure ASMs. ASMs are a measure of the airline's total capacity available for sale. Load factor is RPMs as a percent of ASMs,

Figure 7
Boeing Advanced Model 727-200

and tells how well they are filling their planes with passengers. When load factors are above seventy percent, the airline is doing a good job selling seats.

It is easy to fill an aircraft by lowering fares. An airline could probably have a 100% load factor if it sold every seat between New York and Los Angeles for ten dollars. You would have crowded planes, but it would not be profitable, and soon the airline would be out of business because it couldn't pay all the bills. That is when the profitability terms are considered: yield and cost per ASM. Yield is the measure of revenue collected for each passenger mile. If you paid $180 for 1,000 miles, the airline's yield would be eighteen cents—you paid eighteen cents for each mile your flew. Most U.S. airlines have yield of about fourteen cents. Airlines manage expenses by measuring cost per ASM, which is an airline's total expenses divided by total ASMs. Cost per ASM is reasonable at nine cents. Notice that revenues are usually measured based on passengers, but costs are measured in total seats. So, to determine an airline's profitability, one of the factors needs to be adjusted for load factor.

Major airlines tend to design their schedule for the business traveler, which makes up forty to sixty percent of their passengers, but a far larger share of their revenue. This is because the business traveler is usually not paying the fare from their own pocket, and probably makes the reservation within a few days of the flight. Since the business traveler is less interested in the fare than in a convenient schedule, they usually pay a higher fare, and airlines design their schedules to appeal to them. Airlines can forecast how many of these high-paying passengers will be on a flight based on historical averages, and they offer the rest of the seats at lower fares to vacationers—usually in highly advertised fare sales. As noted above, the airline will already incur most of the expenses for the flight, and by having seats for high-paying business travelers and filling the rest with vacationers at lower fares, the airline maximizes its revenue for the flight. Obviously, the anticipated mix of travelers, and the fares they are willing to pay, will vary based on the cities involved, day of the week, time of day, and season of the year. The airlines try to adjust for all of these factors to maximize their profits for a flight.

Financially, it is also important to note that there are significant exceptions to averages. These averages are based on the largest U.S. airlines: American, United, Continental, Delta, and Northwest. A notable exception to these is Southwest, which averages lower fares and lower costs per seat mile. On the other end is U.S. Airways, which can charge higher than average fares because it operates primarily in the Northeastern U.S. where it has been less exposed to low-fare airline competition. However, costs per ASM are also above average, which has become a problem as airlines like Southwest have moved into the Northeast. Foreign airlines also tend to have higher fares and costs because they have less competition than U.S. airlines.

Hubs are airports where airlines funnel passengers from many cities on to other cities with connecting flights. While it may seem inconvenient to connect through Atlanta, Minneapolis, Pittsburgh, or one of several other cities, the hub-and-spoke structure benefits people in many communities which could otherwise have less or no air service. For example, there are not enough passengers to justify flights such as Tucson to Orlando, Colorado Springs to Raleigh/Durham, or Sacramento to Philadelphia. But, by making all these cities spokes on a hub at Dallas/Ft. Worth, American can economically give a person in Tucson several daily choices for a flight to DFW, with a connection to Orlando, Raleigh/Durham, Philadelphia, or dozens of other cities. Certainly, air service from Tucson to each of these cities would not be economical by itself, but by grouping all of the passengers from Tucson on a flight to DFW, where they are grouped with people from other cities who need to fly to Orlando, enough people are gathered to justify a few flights every day to many cities which otherwise would not have jet air service.

These hubs, and the airlines that operate them, compete against each other for connecting passengers. To use the Sacramento to Philadelphia example, the passenger could connect through Houston on Continental, Chicago on United, Dallas/Ft.Worth on American, Phoenix on America West, Salt Lake City on Delta, or Minneapolis on Northwest. While there may not be much competition on a certain non-stop route, there are usually several airlines serving a given pair of cities through a choice of hub airports. Airlines usually do charge a higher fare for non-stop service, but fares are very competitive on connecting flights because the competing airlines offer alternatives, though they may be via different cities.

With these large networks of routes through several hubs, airlines try to keep a passenger on their own flights as much as possible. Hubs allow major airlines to cover virtually the entire country, and coupled with frequent-flier miles, earn the loyalty of passengers.

Airlines also attempt to dominate routes and cities. Experience has taught the airlines that the largest airline in a given city or a given route will earn more than its proportionate share of the passengers. For example, if airline A has sixty percent of the flights from one airport to another, it will get more than sixty percent of the passengers on that route. This is because they are recognized as having the most convenient schedule, and people will usually think of them first when making travel plans. Conversely, the second-place airline will not get its fair share of passengers. This is why airlines, in recent years, have retreated to cities where they have a large presence, though they still serve all the other spokes from their hubs to pull passengers onto their flights instead of someone else's. Thus, Continental almost pulled out of Denver, leaving it to United, which promptly raised fares. Hubs dominated by one airline will have higher fares for travel from that city due to lack of competition. Ask travelers in Atlanta about Delta's fares, or travelers between Philadelphia and Pittsburgh who have only U.S. Airways to fly on.

In recent years, this network-building by airlines has gone beyond the United States. U.S. airlines have generally been the most aggressive at designed global networks, by tying their flights with international partners to offer an integrated network with convenient connections, common pricing, and combined frequent-flier programs. This allows airlines to extend their reach without having to purchase additional aircraft and find facilities, not to mention the rights to fly international routes, which are usually governed by treaties between the countries' governments. This also allows airlines to dominate international pairs of cities and extend the logic behind the domestic hub-and-spoke networks to international flights. Now, a passenger traveling from Denver to Tel Aviv can connect through London, New York, Brussels, Amsterdam, Zurich, or Frankfurt. Each of these connections will keep the passenger within an airline alliance, so they can get all of their boarding passes and check their luggage through Chicago, even though different airlines are included in the itinerary.

Each of these tactics is designed to attract every additional passenger to a single airline and collect their fares. Since each dollar counts to make a flight and airline profitable, airlines will continue to find new ways to get passengers onto their planes. The example we have seen in the United States is a model for what can be expected in international markets in the future, as competition is allowed and airlines become more innovative to capture revenues and profits.

24

Security for Airplanes and Airports

Terrorism has become an accepted method of achieving political goals by many groups around the world. Unfortunately one of the easiest and most publicized targets is an airplane full of innocent passengers and crew members. Airline passengers are feeling more relaxed recently due to both increased efficiency of security screening at airports, and little publicity when an incident does occur. Although the United States has seen a downward trend in hijacking attempts and airplane bombings in recent years, incidents still occur at foreign destinations. Using 1996 as the most recent year with full statistics available, the number of reported criminal acts against civil aviation worldwide had increased nearly 37% over 1995.

Asia had the distinction among the seven geographic regions of having the greatest number of attacks against civil aviation during 1996. Ten incidents were recorded, two more than in the previous year. Among the incidents were four hijackings, two airport attacks, and one attempted bombing of an aircraft. Six incidents were recorded in Europe: three hijackings, two airport attacks, and one off-airport attack. Two incidents each took place in Spain and Turkey, and one incident each in both Malta and Cyprus. An Iberia flight to Cuba was diverted to the United States, and a Cyprus Turkish Airlines flight to Turkey was diverted to Germany. An Aeroflot flight from Moscow to Nigeria was hijacked following a stopover in Malta.

Latin America and the Caribbean had the second highest number of incidents, with the most involving Cubans seeking asylum. The Middle East and North African regions are also having increased incidents. Looking at trends over the past five years, recorded incidents worldwide have dropped by two-thirds, but aviation related attacks are still very much a way for terrorist groups to make a statement by intimidating the traveling airline passengers so the world press will publish their political motives.

Security screening has been the most successful deterrent to individuals attempting to make a statement by using an airplane to put forward their ideas. Any airline passenger has a story involving a security screener, especially when the passenger had a close connection and their lap top computer had to be checked out before they could go to the airplane. So who are these screeners? They have a 500% turnover rate, very little training, and are paid minimum wage: the airline passenger is getting what is being paid for. We know the screeners can prevent airline incidents if they do their job correctly, but we also know that low wage workers do not really care too much about being a professional life-long airport security screener.

New technology for screening both checked luggage and carry on items can better detect things that should not go on airplanes. But it remains the ability of the person running the machine to detect these items which determines if they are effective. Test articles are run through the check points frequently to check the screeners' ability to detect illegal items. Experienced screening personnel can considerably decrease the amount of time passengers are in line at the screening point. When the line moves quickly, it ultimately saves the airline time and money.

Points to remember are that anyone carrying a laptop computer or any large electronic equipment should expect to open it up for inspection and power it up. If the screener working the x-ray machine calls for a bag check, no one can override that decision. All checkpoints will have a supervisor who you can ask for, or you can contact the airline's ground security coordinator. At the beginning of each shift, a supervisor tests the walk-through detectors with a test device specifically for that purpose. A walk-through metal detector that fails cannot be used. Screeners are taught to look for anything that is a weapon, or anything that could hide or mask a weapon.

Explosive Trace Detection (ETD) Equipment is being installed at many airports. This equipment uses technology that analyzes the size, density, and physical characteristics of objects in bags and compares those to the known composition of explosives to determine if a threat exists in the bag. Explosive Detection Systems (EDS) and other enhanced x-ray systems have been purchased by the FAA and given to airlines to be used at major airports.

International flights will be treated more intensively for possible threats than those flights remaining in the United States. All aircraft with a U.S. origination point destined for an international location will have the cargo hold searched prior to the loading of any freight or baggage. A search of the airplane cabin will be made for items that do not belong. The search area includes overhead bins, closets, lavatories, galleys, trash receptacles, storage bins, underneath seats, and other readily accessible compartments in the passenger cabin.

During the 1970s, the government started the Federal Air Marshall program as an initial response to aviation related criminal acts, which were increasing at the time. Federal agents disguised as normal passengers were boarded on randomly selected flights. This program was reactivated in 1985 with the hijacking of TWA Flight 847. In 1998 this program is again to be geared up, for use mainly on international flights.

Additionally, passengers will see baggage being matched with the actual passenger to prevent luggage being boarded that does not match up with someone actually on the airplane. The baggage of randomly selected passengers and passengers meeting FAA profile criteria will be marked with special tags. Ramp personnel will not load baggage marked with these tags until the gate agent verifies that the owner of this baggage has boarded the aircraft. As an airline passenger, please understand that security screening is not meant to harass or intimidate you, but is simply a logical response to individuals who want to interfere with normal airline travel by making threats in an attempt to get the press to tell the world about their special problems. The airlines, along with the FAA, are getting better at dealing with this problem; there will be fewer delays in the future and the system will become more efficient.

25

Commuter Airlines

Deregulation of the airline industry led to many changes, of which the expansion of the commuter airline industry was probably the biggest. Many entrepreneurs realized the profitability of flying smaller airplanes on short routes, most of which had been abandoned by airlines which were changing their fleets to large, long-range jets. While the major airlines had been strongly regulated by Federal Aviation Administration rules, airplanes with less than fifty seats were subject to less cumbersome rules and regulations, until recent changes were mandated by the FAA. Making the commuters subject to the same rules as larger airlines was probably due to the fact that commuter airplanes were involved in more accidents than larger airplanes.

Costs of operating a commuter airplane with two smaller engines were significantly less than operating a large airplane, like those produced in the 1960s. Regulatory paperwork requirements alone were much less. Maintenance costs were much lower for smaller, simpler, turbo-prop engines, than the larger fan-jet engines needed for the bigger jets. Pay for pilots was much less, since many pilots hired by commuters had not accumulated enough flight time to be hired by major airlines. Pilots could fly longer hours with less rest requirements than were allowed for the larger airlines, making pilot productivity much better. Additionally, FAA rules for airplanes with less than fifty seats required only one flight attendant. Meals are not served and not many cokes or coffees are served on a short flight, reducing airline costs even more. Co-pilots usually help load baggage and freight, and almost all employees work several jobs on non-union pay scales. Costs for employee benefits and pensions are also very low, due to the fast employee changeover.

Some of these conditions led to an increased accident rate when compared to the major airlines. But in fairness, it must be pointed out that most accidents happen on take-off or landing, and by subjecting pilots to many more flight segments, the odds of an accident were sharply increased. New FAA rules will hopefully see this trend become comparable to other air transportation services.

Originally, commuter airlines were flying smaller airplanes powered by turbo-prop engines, which are jet engines with a propeller on the front for thrust, rather than thrust developed from internally developed combustion. In the past few years, some foreign airplane producers have started making commuter airplanes that have no propeller, but are turbo-jet powered. New, smaller jet engines, with low fuel consumption and ease of maintenance, are making the new sleek commuter jets very desirable for smaller airlines. Propellers disappeared from large airlines about twenty years ago, and now within the next few years, propellers will completely disappear from commercial airline transportation.

Most commuter airlines have agreements with major airlines for joint ticketing and reservation services. Ownership of commuter airlines is also changing, as major airlines purchase the smaller commuters which feed passengers into the hub cities for flights to other major cities. This is causing labor problems, as pilot unions attempt to keep pay scales high. Pilots of the commuters want to be integrated into the seniority lists, to insure a job with the big airlines. Other innovative ideas are being tried to make everybody happy, such as "B" pay scales, where new employees are paid less than older employees, as the airlines attempt to keep costs low. Naturally, the unions get excited about this issue.

As stated previously, controlling a hub city has proven to be very economically advantageous for the major airlines. Therefore, having their own connecting commuter flights makes hub city dominance even better. A commuter airline can be started with little capital for a few airplanes to serve a limited market. With hub city dominance being so important, the major airlines will continue taking over the commuter airlines, either by directly buying them or having strong code share arrangements, insuring that passengers transiting a large city will stay on one airline.

Southwest Airlines is the one airline that has risen from commuter status to a major airline, originally serving only the Dallas/Houston market with three airplanes in 1975. Having a bit of an advantage by using Love Field in downtown Dallas, when all other airlines moved to the new Dallas/Ft. Worth regional airport, Southwest continued to expand over Texas, and now flies over the entire country except for New England. Almost all other commuter airlines are now part of larger airlines.

26

Pilot Stuff

I have often told people that I have never worked a day in my life, because I love being a pilot so much that it has never been work for me. Although there is tremendous responsibility associated with the job title of airline pilot, and the flying time can be very demanding upon occasion, most pilots I know enjoy their job so much that it doesn't feel as if they "work" for a living. Some have actually been quoted as saying they would pay to get a flying job with a major airline. Of course, there are some who would complain about anything, just like any other job.

Most pilots have dreamed of flying since seeing their first airplane at a very early age. Getting into the cockpit as a airline pilot takes lots of flying hours, which can be accumulated either as a military pilot, or the long way, with only civilian flight time. Many have started by hanging around airports, begging flying time for pumping gas or cleaning airplanes. Others have completed a college degree to become eligible for flight training in the armed forces.

Building flight time by the civilian route requires a lot of hard work over a long period of years. After taking very expensive lessons, one can hopefully solo a little airplane within ten hours. After many hours of practice, including cross-country navigation and instrument training, the new pilot can take the tests for a commercial rating. This rating allows the pilot to start making money while increasing his flying hours, by getting a job as a co-pilot for a corporate flying team. Another choice would be to get an instructor rating and start teaching new pilots how to do what he has learned. Neither of these situations pays very well, so most will be doing this as a second job, mainly to build up the flight time required for employment by the major airlines. An FAA certified log-book is used to record a pilot's hours of flight time.

The United States Armed Forces also train pilots. The long hours and expenses are paid by the U.S. taxpayer in exchange for eight years of service to help protect the United States. After winning his military wings, the government convinces him that he is the best pilot in the world, and says you belong to us for the next eight years. Flying a supersonic jet with magic ordnance on the wings over hostile territory, and then returning to a country club, can be a tremendous experience for those who are fortunate enough to have done it. In the long run, for most, the bureaucracy of working for the government, combined with better pay, sends one in search of airline job applications within a few years.

Just as the price of fuel can make logging light airplane time expensive, a war changes the availability of military pilots, and the economy determines whether the airlines are hiring. A lot of the game is timing. If you can build up your time and be available when the airlines are buying new airplanes, or getting rid of the pilots over sixty years old, then you can start flying in clean cockpits with meal service and a different hotel every night.

Since the major airlines generally require about 2,000 hours of flight time and at least a commercial pilot rating, another way of getting hired with a big airline is to spend time as a commuter pilot. As mentioned before, commuter airline's pay scales are very low, and pilots fly a lot more short flights.

Entrance requirements are usually much less and the turnover rate is much higher, due to the increased work load and decreased salary scales. But many pilots do work under these conditions simply to build up time until the major airlines are hiring.

Most new airliners today only have two pilots, the captain and the first officer, or co-pilot. Older airplanes, and some new ones, also have a third crew member called a second officer, or flight engineer. The flight engineer position is a tradition dating back to the original propeller-powered airplanes, when the mechanic went along with the two pilots to operate the airplane systems and fix the plane when it broke. Pilot unions dictated the three-man crews until very recently, but the economy of paying two pilots instead of three made the airlines start buying airplanes with only two pilot positions. Also, technology produced better and safer airplane systems not requiring a system manager.

In the past, therefore, a new hire pilot was hired as a flight engineer, whose job was to do all the paperwork for the flight. This included the exterior airplane check, being sure the correct amount of fuel was on board, and turning on the galley power so the captain could have his coffee. Things haven't changed much since the mechanic went along on the Lockheed Constellations and the DC-6s. Today, most new pilots will get directly into the right seat as a co-pilot, and will actually fly every other take-off and landing. In airline terminology, one flight segment is a leg, therefore, the captain and co-pilot alternate flying legs of a trip.

The actual airline training for a new pilot initially consists of about three weeks of ground school to learn the systems of the airplane he will be flying. Following this, a few days are spent in CPTs or cockpit procedures trainers, where the familiarity with the cockpit is started, including checklist usage and basic procedures. This is followed by about ten simulator "rides" in which the procedures of actually flying are duplicated in a real cockpit, less the remainder of the airplane, that actually moves and supposedly reacts like a real airplane.

Although simulators have become progressively better, it still remains that you are simply operating a video machine just like the ones you have to put quarters in at the arcade in the mall. This "hands on" training is augmented by both written and oral examinations about the airplane and procedures. Following completion of all training, the new pilot will fly an actual flight with a check-pilot, or a pilot instructor certified by the FAA to "sign-off" new pilots as being qualified to start "flying the line." A new pilot's first flight in the airplane will be with passengers in the seats behind him.

Seniority is probably the most important aspect of a pilot's career, and it starts the first day on the job for the airline and never changes. Seniority cannot be transferred to another airline should the pilot elect to change employers. If an airline goes out of business, it is time to start all over again. Pilots bid for trips on a monthly basis according to seniority, and the less senior pilots often end up on call as reserve pilots. A monthly schedule of trips is published for guaranteed flying days and pay. A reserve pilot will be on call for a trip for most of the month, with ten days off, while a senior pilot who is awarded a line, will only fly about fifteen days per month, with the rest of the days off. Seniority, based on date of hire, coupled with the airline pilot requirements, determines when a pilot is eligible to upgrade to co-pilot or captain. Monthly bidding for the best trips or lay-over cities are based only on seniority, with the most senior pilot getting first choice and the pilots lowest on the list getting to stand by on reserve. When upgrading to a new status or different airplane, another bidding process starts, again based on how long the pilot has been employed at the airline.

As the airline industry began growing significantly following World War II, pilots in the newly formed Air Line Pilot's Association (ALPA) visualized larger airlines needed to fly the many new passengers and freight in the nation's booming economy. The union bargained for a pilot pay scale based on the weight of the airplanes being flown. I doubt there was one airline president or pilot, for that matter, who ever dreamed of an airplane the size of the Boeing 747. Pilot pay scales are based on the weight of the airplanes they fly. Therefore, the lowest-paid pilots fly little twin jets such as the DC-9, while the older, more senior pilots, fly heavy airplanes such as the 747s, and receive a much larger paycheck. This pay schedule exists for all pilots that belong to ALPA, which includes all major U.S. airlines except American, whose pilots are members of the Allied Pilot's Association, and Southwest and Continental, whose pilots are represented by independent associations.

Training for airline pilots continues throughout their career. Every time they bid to fly a new airplane, the pilot is required to complete the entire ground-school training program, including an extensive simulator program. Additionally, every six months a captain is required to complete a proficiency check of all procedures in a simulator training period. Ground schools for reviewing company policy and FAA rules are also required to be attended twice each year. Actual check rides in the airplane by FAA certified pilots are also accomplished annually.

Medical regulations are continually monitored by the FAA. Captains are required to pass two physical examinations each year and co-pilots must pass one exam—all given by FAA-designated doctors. Additionally, pilots are required to report any medical changes, drug use, DWI convictions, or alcohol abuse.

Financial rewards for flying jet airplanes are significant for most pilots. Although initial pay scales, especially at commuter airlines, are very low for the responsibility involved, the pay scales appreciate to a point where a pilot with high-seniority, flying for only fifteen days a month, can be very rewarding. A captain on a small two-engine jet such as a Boeing 737 can have an annual income of over $100,000, while the older guys flying heavy jets such as the 747 will have annual incomes approaching $250,000. Co-pilots receive about sixty percent of the captain's pay and flight engineers receive fifty percent, all based on years of service. Airlines all have different compensation packages based mainly on the union's bargaining ability, but the pay scales for the six larger airlines are just about the same, with variations for profit sharing or company stock ownership. Pensions are equally well-funded, with retiring pilots receiving about half of their last salary as retirement pay.

Historically we all know the captain of a ship was the unquestioned officer in charge. This tradition continued when airplanes began flying with a captain in charge, and the idea of questioning a captain's decision was unthinkable. This attitude continued in cockpits until very recently, when research showed that a captain could indeed make a mistake in judgment. Some incidents were directly attributable to a young co-pilot being afraid to question a senior captain. Fortunately, this is being changed. Airlines are now requiring pilots to attend seminars involving crew resource management, which educate all cockpit crew members on how to communicate with each other more successfully. This theory teaches old captains to listen to other crew members, and also teaches new pilots to question the captain's authority if they feel strongly enough about what is happening in a given situation. Interestingly, this training is now called "error management" and is completely endorsed by the Federal Aviation Administration. Obviously, the acknowledgment by airline management, pilots, and the FAA that change is for the better, will benefit us all.

27

History of Air Transportation

The first air passenger services began in 1910, when dirigibles began operating between several German cities. The first scheduled airplane service to carry passengers began in the U.S. in 1914. Several experimental airmail flights took place in India, Europe, and the United States before World War I, but sustained air transport services were not firmly established until after the war. In 1918, the U.S. Postal Department began the nation's first scheduled air service, using planes and pilots furnished by the War Department. By the mid 1920s, there were regularly scheduled airmail flights from coast to coast. On night flights, pilots were guided by flashing beacons at ground stations along the flight path.

With the passage of the Air Mail (Kelly) Act of 1925, private lines began carrying mail under contract in single-engine, open-cockpit planes. Another federal law, the Air Commerce Act of 1926, established government agencies for the development of airports, radio navigation, and other aviation related services. Although several independent companies had begun passenger service early in the 1920s, none had succeeded. With subsidies from the Air Mail Act, however, new airlines like American, Delta, and United began to see their operations become profitable. These airlines soon were flying larger monoplanes with enclosed cabins, such as the Ford Trimotor, which carried up to fifteen passengers.

In Europe, many governments developed an extensive airline system. Although early airmail routes compared unfavorably with American round-the-clock deliveries, European passenger operations became much more sophisticated. By 1929, Britain was operating a commercial air route to India, and within a few years, other nations began combined mail, freight, and passenger service to distance countries and dependencies.

The era between the world wars saw significant advances in weather forecasting, navigation equipment, aerodynamics, and innovative management. Symbolizing these trends were airline specifications for modern airplanes, such as the DC-3 of the 1930s, a streamlined metal monoplane with reliable and efficient radial engines, retractable landing gear, and variable-pitch propellers Pan American Airlines bought larger airplanes capable of landing on water, and began to link Europe and Asia with the United States. (see Figure 8)

During the 1940s, intercontinental air transport became firmly established. After the war, the new long-range, four-engine transports with pressurized cabins and advanced instrumentation were increasingly able to avoid bad weather systems, enhancing passenger comfort and making operations more economical and reliable. These new planes, and the jet airliners introduced in 1958, replaced railroad trains and ocean liners as the primary mode of long-distance travel. A new generation of wide-bodied "jumbo jet" transports began operating in 1970, and the Concorde started supersonic passenger flights in 1976.

Air transport operations are monitored and regulated by several national and international bodies. Within the U.S., the Air Commerce Act of 1926 charged the Commerce Department with encouraging civil aviation, as well as setting standards for both airplanes and pilots. In 1938, these duties

Figure 8
Boeing Model 314A Clipper

were transferred to a new Civil Aeronautics Authority, which was then split in 1940 into two agencies: the Civil Aeronautics Board (CAB), and the Civil Aeronautics Administration (CAA). The CAB was entrusted with economic and safety rule-making, as well as accident investigation; the CAA with airway development and enforcement of safety standards. In 1958, Congress gave the new Federal Aviation Agency (FAA), responsibility for rule-making and enforcement in the safety field, and for maintaining a common civil-military system of air navigation and air traffic control. In 1967, the FAA was renamed the Federal Aviation Administration and placed within the newly created Department of Transportation. At the same time, accident investigation was transferred from the CAB to the new National Transportation Safety Board (NTSB). Legislation passed in 1978 phased out the CAB's remaining function, airline economic regulation, and in 1984 that agency ceased to exist. The International Civil Aviation Organization (ICAO), was created as a permanent body of the United Nations in 1947. Since then, it has facilitated the establishment of worldwide standards for safety, reliability, and navigation, while assisting in resolving legal issues.

Modern airline operations include a significant effort in other services, such as airframe and engine maintenance; personnel training of pilots, flight attendants, tickets agents, and ground crews; maintenance of computerized reservation and accounting equipment; food preparation; and the operation of hotels and resorts around the world. Most of the hotels and resorts have since been sold in an effort at consolidation. To meet federal standards for training and flight proficiency requirements, airlines rely heavily on computerized simulators, because exclusive training in complex jet transports is far too costly and time consuming.

Along with airmail service, the shipping of commodities by air began in the 1920s. The private airline of American automobile manufacturer Henry Ford carried considerable amounts of bulky freight in the late 1920s, and one or two U.S. firms offered all-cargo schedules in the Great Lakes area about the same time. The air cargo business in the 1930s expanded as larger airliners came into service, but air cargo still lagged far behind passengers and mail as a source of revenue. Nonetheless, for certain categories of compact, lightweight, and high-value items, air transport proved highly useful. Typical air cargo in the pre-World War II era included jewelry, cut flowers, high-fashion clothing, movie reels, pharmaceuticals, and high-priority replacement and repair parts for machinery. Even some live animals were carried.

After World War II, when larger, more efficient lanes became available, airlines increased their cargo operations, and several all-cargo companies were formed. Postwar freight was likely to include almost anything that would fit inside the fuselage of an airplane. During the Berlin airlift of 1948-49, military planes carried not only food, but also tons of coal and construction equipment. With increasing frequency, all kinds of livestock—from horses to zoo animals—traveled in pressurized cargo planes. Perishable produce, frozen foods, heavy machinery, automobile parts, and even complete automobiles, were also characteristic of air cargo. Smaller shipments were usually packed in large containers to make for easy handling and to cut down on pilferage. In the 1970s, the overnight delivery of letters, packages, and even larger cargo became of increasing importance, both within the U.S. and internationally. New carriers were now competing with the national postal service.

Compared with truck, rail, and water transport, air cargo costs are still higher in cost per mile, and air cargo still commands only a small share of total inter-city tonnage moved by all forms of transport in the U.S. Nevertheless, many shippers still profit from air cargo by reducing inventories, warehouse requirements, and handling charges.

As a result of the 1978 law deregulating the air transport industry, many airlines began to drop less profitable service to smaller cities, formerly required by law, and began intense competition in high-density passenger markets. New carriers, including the low-fare People Express, began flying. The major airlines adopted a hub-and-spoke system, in which each carrier funnels its routes into a handful of major regional airports. This scheme maximized the profitable use of an airline's fleet, but it also eliminated scores of direct flights between smaller cities. Fare wars, that benefited passengers at first, created a profit squeeze for a number of carriers hard hit by rising costs. In the mid and late 1980s there were several mergers, acquisitions, and bankruptcies. For example, in 1987, the undercapitalized People Express, which had merged with Frontier, was absorbed by Continental Airlines, which then in 1990 filed for bankruptcy while continuing to fly. TWA did the same in 1992. Eastern and Pan Am, both of which had been major factors in the airline industry since their founding in the 1920s, ceased operations in 1991. Many smaller commuter lines, which generally had prospered by taking over the less-busy routes abandoned by the major companies, became subsidiaries of the truck lines that remained in business.

During the 1970s, the number of domestic passengers of U.S. airlines increased about seventy-eight percent, and during the 1980s the figure was up to about fifty-eight percent. In 1990, there were four hundred and twenty-three million domestic passengers and nearly forty-two million international passengers, a seventy-five percent increase over 1980. Total freight and express cargo flown by U.S. airlines almost doubled during the 1980s, from 5.7 billion to 10.6 billion ton-miles in 1990. Total airline profits were down, however, and the industry lost about four billion dollars in 1990. Employment figures for U.S. airlines were up over fifty percent during the 1980s, from about 360,000 to 545,000.

Worldwide, 925 scheduled airlines operated 14,651 aircraft in 1990. In the U.S., the twenty-two scheduled airlines, belonging to the Air Transport Association, operated 4,275 airplanes, including freight aircraft.

As the turn-of-the-century approaches, international alliances between airlines seems to be the trend. All U.S. airlines are starting code-sharing with foreign airlines to make international travel easier for the passenger, and also more lucrative financially for the one airline system that can take them anywhere in the world.

28

Trivia Questions

1. How many hours can an airline pilot fly per year according to the FAA?
2. What is the language used by tower controllers talking to pilots at the Tokyo airport?
3. What is the limiting or maximum speed for airplane tires on the ground?
4. If a pilot squawks "7700" on his transponder, what is he telling the ground controller?
5. How many flight attendants are required on an airplane with 180 passenger seats, but only ten passengers?
6. Airplane identification numbers painted on the airplane side starting with "N" are from what country?
7. What is the temperature outside the airplane window when cruising at 40,000 feet under normal conditions?
8. Will overheated airplane tires explode?
9. What airline does not paint its airplanes, except for the company name?
10. How many tires are on a Boeing 747?

11. What is the three letter symbol for Chicago's O'Hare Airport and why was the name changed?
12. What airports are named for presidents?
13. According to the Department of Transportation, a flight is on time if it arrives at the gate within how many minutes of scheduled arrival time?
14. For flight planning weight and balance numbers, a ten-year-old boy weighs how much?
15. What is the altitude separation requirement for airplanes flying over the North Atlantic?
16. Name the airplanes originally designed for airline use now being used by the armed forces.
17. What seat row number is not used on most airliners?
18. How many engine turbine sections are there on a Boeing 747?
19. How many gallons of fuel would Air Force One use on a flight from Washington to Moscow?
20. May you bring your favorite dog along in the passenger cabin without buying a ticket?
21. Which airport has the most runways?
22. May a pilot fly an approach with reference to ground objects, such as bridges andhighways?
23. You can save money by bringing how many bottles of your own whiskey on board a flight to mix your own drinks?
24. May a pilot use eyeglasses and hearing aids while flying?
25. Will flying through volcanic dust damage a jet airplane?

29

Trivia Answers

1. 1000 hours
2. English
3. 195 knots
4. Emergency
5. Four, or one for every fifty passenger seats, no matter how many actual passengers.
6. United States
7. Minus sixty degrees Fahrenheit, or a change of three degrees per thousand feet.
8. No, internal fuse plugs will melt, allowing the air to escape.
9. American Airlines
10. Eighteen, with four on each of four main gear trucks and two on the nose gear.
11. ORD was for Orchard Park Airport but changed to honor Navy pilot Eddie O'Hare.

12. John Kennedy in New York, George Bush in Houston, Charles DeGaulle in Paris, and Ronald Reagan National Airport in Washington.
13. Fourteen minutes late if arriving fifteen minutes after published arrival time.
14. Eighty-five pounds for children between the ages of two and twelve.
15. One thousand feet
16. Boeing 707 and Douglas DC-10 are used as tankers, the DC-9 is used for transport, and a Boeing 747 is used as Air Force One.
17. Row thirteen
18. Five, one in each engine and one on the APU.
19. Approximately 40,000 gallons
20. Yes, if the container the dog is in will fit under your seat, Fido does not require a ticket.
21. DFW or Dallas/Fort Worth in Texas.
22. Yes, a visual approach can be flown if the pilot also has the runway in sight and uses electronic navigational aids as a back up.
23. Zero
24. Yes, as long as it is noted on his Airman Medical Certificate for the FAA.
25. Severe damage especially to the engines.

30

Glossary, Technical Talk, and Expanded Reading

Bernoulli's Principle

Bernoulli's Principle, in physics, is the concept that as the speed of a moving fluid (liquid or gas) increases, the pressure within that fluid decreases. Originally formulated in 1738 by the Swiss mathematician and physicist Daniel Bernoulli, it states that the total energy in a steadily flowing fluid system is a constant along the flow path. An increase in the fluid's speed must, therefore, be matched by a decrease in its pressure.

The principle covers the flow over surfaces, such as aircraft wings or ships' propellers. A wing is designed to permit air to flow more rapidly over its upper surface than over its lower surface, with a consequently reduced pressure on the top surface as compared to that on the bottom surface. The resulting pressure difference provides the lift that sustains the plane in flight. On a rotating propeller, which is also shaped like an airfoil, the corresponding pressure difference provides the thrust that propels the vessel.

Direction Finder

The radio direction finder was the first navigational aid to come into general aviation use. If the bearings of two transmitters with known locations can be measured, the position of the receiver may then be determined. A modern direction finder consists of a conventional radio receiver with an antenna in the form of a coil of wire, called a loop. The loop antenna has strong directional properties; if it is mounted so that the axis of the loop points directly to a radio station, it will receive no signal from that station; if it is mounted so that the plane of the loop passes through the radio station, it receives a strong signal. In practice, a known station is tuned in, and then the loops are rotated until no signal is heard; this position is called the aural null. The axis of the loops must then point directly toward (or away from), the station; this direction is plotted by the navigator as a line of position.

An automatic direction finder (ADF), has a motor that rotates the loop antenna, keeping the loop always in the null position. The motor also actuates a needle, similar in appearance to a compass needle, that indicates the position of the loop. This so-called radio compass points, not toward north as a compass does, but toward whichever station is tuned in on the loop antenna.

Drag

Drag is the force that tends to retard the motion of the airplane through the air. Most drag is a result of the resistance of the air to objects moving through it. Streamlining the aircraft reduces drag. Induced drag is a result of the lift produced by the wing and therefore, is the penalty exacted for creating lift. Parasite drag is caused by any non-lift producing attachment to the airplane.

FAA

Established as the Federal Aviation Agency by the Federal Aviation Act of 1958, the name was changed to the Federal Aviation Administration when becoming part of the Department of Transportation in 1967. The primary responsibilities of the FAA are to regulate air commerce in order to promote its development, safety, and fulfill the requirements of national defense; promote and develop civil aviation and a national systems of airports; manage navigable airspace within the United States and regulate air traffic in the interests of safety and efficiency; consolidate research and development and the installation and operation of air navigation facilities; and develop and operate air traffic control and navigation system for both civil and military aircraft. Can all this be done better by a bureaucracy or a non-government organization?

Latitude and Longitude

A system of geometrical coordinates used in designating the location of places on the surface of the earth. Latitude, which gives the location of a place north or south of the equator, is expressed angular measurements ranging from zero degrees at the equator to ninety degrees at the poles. Longitude, the location of a place east or west of a north-south line called the prime meridian, is measured in angles ranging from zero degrees at the prime meridian to 180 degrees at the international date line.

Midway between the poles, the equator, a great circle, divides the earth into northern and southern hemispheres. Parallel to the equator, and north and south of it, are a succession of imaginary circles that become smaller the closer they get to the poles. This series of east/west circles, known as the parallels of latitude, is crossed at right angles by a series of half-circles extending north and south from one pole to the other, called the meridians of longitude. Although the equator was an obvious choice as the prime parallel, being the largest, no one meridian was uniquely qualified as prime. Until a single prime meridian could be agreed upon, each nation was free to choose its own, with the result that many nineteenth century maps lacked a standardized grid. The problem was resolved in 1884, when an international prime meridian passing through London's Greenwich Observatory was officially designated.

Degrees of latitude are equally spaced, but the slight flattening at the poles causes the length of a degree of latitude to vary from 68.7 miles at the equator to 69.41 miles at the poles. At the equator, meridians of longitude one degree apart are separated by a distance of 69.17 miles, while at the poles, the meridians converge. Each degree of latitude and longitude is divided in sixty minutes, and each minute divided into sixty seconds, thereby allowing the assignment of a precise numerical location to any place on earth.

Lift

Wing area influences lift; the more of the wing that is exposed to the air, the greater the lift. The up or down tilt of the wing, called angle of attack, contributes to or detracts from lift. As a wing is tilted upward, that is, as its angle of attack is increased, its lift increases. The air passing over the top of an uptilted wing must travel a greater distance and thus produces a greater pressure differential between the upper and lower surfaces.

Aircraft speed has a great influence on lift. The faster the air moves over and under the surfaces of an airplane, the greater the pressure differential and, as a result, the greater the lift.

As an airplane flies on a level course, the lift contributed by the wing and other parts of the structure counter-balance the weight of the plane. Within certain limits, if the angle of attack is increased while the speed remains constant, the plane will rise. If the angle of attack is decreased, that is, if the wing is tilted downward, the plane will lose lift if speed is constant. Lift varies directly with speed.

During the course of a flight, the pilot frequently alters the speed and angle of attack of the aircraft. These two factors are often balanced against each other. For instance, if the pilot wishes to increase speed and yet maintain a level flight, the angle of attack must be decreased to offset the extra lift that is provided by the increase in the speed of the aircraft.

In preparing to land, the pilot must ease the plane down and at the same time reduce its speed as much as possible. To compensate for the considerable loss of lift resulting from the decrease in speed, the pilot provides lift by altering the wing area, effective curvature, and angle of attack. This is done through the use of flaps—large wing extensions located at each trailing edge. Most flaps are normally tucked into the wing during straight and level flight. If extra lift is wanted, the pilot extends the flaps out and down.

Airplane designers try to design aircraft with high lift-to-drag ratios, that is, much more lift than drag. They are limited, however, by factors such as speed and the weight that the plane must carry. Faster planes usually have lower ED-to-drag ratios. A subsonic modern transport has a lift-to-drag of about fifteen and a light private plane may have a lift-to-drag ratio of about twenty-five. Supersonic transports have a ratio of about six.

Loran

Long Range Navigation, a system developed in World War II, allows a navigator to establish a position by finding the difference in the time that it takes radio signals to reach them from two synchronized transmitters spaced some distance apart. The loran transmitter system consists of a master and a slave station. The master station emits a short pulse or signal at regular intervals, and this pulse is repeated by the slave station, which is controlled from the master station by radio. Both of these signals are received aboard the aircraft and are recorded as small sine waves on a cathode-ray tube. Because radio waves travel at a uniform speed, the location of all points where the signals from the two station are separated by a given interval, can be represented by a definite curve called a hyperbola. By switching to another pair of loran transmitters and repeating the procedure, the navigator can find another curve representing the airplane position, and can thereby determine the actual position at the intersection of the two loran lines of position. This is called a fix.

Mach

Mach number, the ratio of the speed of an airplane to the speed of sound, was named for Ernest Mach, an Austrian physicist, born in Czechoslovakia and educated at the

University of Vienna. His outstanding work in ballistics contributed greatly to the theory of flight.

An airplane traveling at less than Mach One is traveling at subsonic speeds; at about Mach One, transonic, or approximately the speed of sound; and greater than Mach One, supersonic speeds. An airplane flying at Mach Two is going twice the speed of sound.

NTSB

The National Transportation Safety Board is an independent agency of the United States federal government, created by an Act of Congress in 1975, which promotes transportation safety by investigating accidents, locating problems, and making recommendations on safety measures to industry and government agencies. The board's functions include: determining the causes of all U.S. civil aviation accidents; railroad accidents involving a passenger train or any fatality; highway casualties; and major marine accidents. In addition, the board is responsible for establishing regulatory requirements for reporting accidents and assessing the procedures for the transportation of hazardous materials. Headquartered in Washington, the NTSB has five members appointed by the president for five-year terms.

VOR

VHF Omnirange is a radio range with an infinite number of beams. The omnirange station has four antennas; the central antenna broadcasts a continuous reference signal while the others broadcast a variable signal. By measuring the phase difference between the antennas the receiver can determine its bearing from the station. In practice, the omnirange receiver has three modes, one which can be set manually to any desired course, a second which tells whether the plane is left or right of the course, and a third which resolves 180 degree ambiguity by indicating "to" or "from" the station.

About the Author

Captain Larry Coy was born in Pennsylvania and graduated from Shippensburg High School in 1958. He received his B.S. in mathematics education from Shippensburg State College in 1961. In December 1961, Coy began his U.S. Navy Flight Training at Pensacola and was awarded Navy Wings in May 1963. Assigned to VQ-1 at Atsugi, Japan, Coy flew EC-121 Constellations and has flown throughout the Pacific with extensive time on deployment in Vietnam.

In November 1966, Coy left the Navy and joined Continental Airlines. Flight Engineer on a Boeing 707 was Coy's initial assignment, although most of his energy has been spent flying a Boeing 727 as a Second Officer, Co-Pilot and Captain. He has been Airline Transport Pilot rated on Boeing 727, Boeing 737, Airbus A-300, and Douglas DC-10.

Coy now holds over 25,000 flying hours beginning with a Piper Cub solo at age sixteen and ending as a Captain on DC-10 flying from Newark to Europe.

It is Captain Coy's sincerest hope that this book will share his love of flying with the passenger's in the cabin of his airplane.

Please forward any comments or questions regarding airline flying for use in future editions to:

Passenger's Guide
3952 Highway 150 West
Box 727
Mooresville, North Carolina
28117
USA